Sediment Microbiology

Special Publications of the Society for General Microbiology

PUBLICATIONS OFFICER: COLIN RATLEDGE

1. Coryneform Bacteria,
   eds, I. J. Bousfield & A. G. Callely

2. Adhesion of Microorganisms to Surfaces,
   eds. D. C. Ellwood, J. Melling & P. Rutter

3. Microbial Polysaccharides and Polysaccharases,
   eds. R. C. W. Berkeley, G. W. Gooday & D. C. Ellwood

4. The Aerobic Endospore-forming Bacteria:
   Classification and Identification,
   eds. R. C. W. Berkeley & M. Goodfellow

5. Mixed Culture Fermentations,
   eds. M. E. Bushell & J. H. Slater

6. Bioactive Microbial Products:
   Search and Discovery,
   eds. J. D. Bu'Lock, L. J. Nisbet & D. J. Winstanley

7. Sediment Microbiology,
   eds. D. B. Nedwell & C. M. Brown

Sediment Microbiology

Edited by

D. B. NEDWELL

*Department of Biology,
University of Essex,
Colchester, Essex, UK*

and

C. M. BROWN

*Department of Brewing
and Biological Sciences,
Heriot-Watt University,
Edinburgh, UK*

1982

Published for the

Society for General Microbiology

by

ACADEMIC PRESS

A Subsidiary of Harcourt Brace Jovanovich, Publishers

London New York

Paris San Diego San Francisco São Paulo

Sydney Tokyo Toronto

ACADEMIC PRESS INC. (LONDON) LTD.
24/28 Oval Road
London NW1

United States Edition published by
ACADEMIC PRESS INC.
111 Fifth Avenue
New York, New York 10003

British Library Cataloguing in Publication Data
Sediment microbiology.—(Special publications of the
Society for General Microbiology; no. 7)
1. Aquatic microbiology—Congresses
I. Nedwell, D. B. II. Brown, C. M. III. Series
576'.192'083 QR105

ISBN 0-12-515380-5

LCCCN 81-71581

Printed in Great Britiain by
Whitstable Litho Ltd., Whitstable, Kent.

CONTRIBUTORS

BATTERSBY, N.S. *Department of Brewing and Biological Sciences, Heriot-Watt University, Chambers Street, Edinburgh EH1 1HX, UK.*

BILLEN, G. *Laboratory of Oceanography, University of Brussels, 50 Avenue F.D. Roosevelt, 1050 Brussels, Belgium.*

BROWN, C.M. *Department of Brewing and Biological Sciences, Heriot-Watt University, Chambers Street, Edinburgh EH1 1HX, UK.*

FRY, J.C. *Department of Applied Biology, University of Wales Institute of Science and Technology, King Edward VII Avenue, Cardiff CF1 3NU, UK.*

HERBERT, R.A. *Department of Biological Sciences, University of Dundee, Dundee DD1 4HN, UK.*

JONES, J.G. *Freshwater Biological Association, The Ferry House, Ambleside LA22 0LP, UK.*

MALCOLM, S.J. *Scottish Marine Biological Association, PO Box 3, Oban, Argyll SP34 4AD, UK.*

MAXWELL, J.R. *Organic Chemistry Unit, School of Chemistry, University of Bristol, Bristol BS8 1TS, UK.*

STANLEY, S.O. *Scottish Marine Biological Association, PO Box 3, Oban, Argyll SP34 4AD, UK.*

WARDROPER, A.M.K. *Organic Geochemistry Unit, School of Chemistry, University of Bristol, Bristol BS8 1TS, UK.*

PREFACE

In general the study of sediments has been a neglected area of aquatic microbiology when compared with the larger volume of information published on microbial processes in the water column. This may be because sediments have historically been considered difficult to sample, handle, and to work with, in comparison to water. This situation has changed dramatically within the last few years, assisted by a realization that sediments play a critical role in the ecology of aquatic systems, particularly in freshwater and coastal marine environments where the water column is relatively shallow. This burgeoning of interest in sediments has been further helped by an increased level of research into the functioning of anaerobic microbial communities; also previously considered to be a difficult area of microbiology in comparison to the more commonly studied aerobic systems. Many sediments are anaerobic, except for a surface aerobic layer of varying depth, and the concurrent development of information on anaerobic communities has reinforced the intrinsic interest in sediment microbiology *per se*.

With the current level of interest in this subject, the Ecology Group of the Society for General Microbiology organized a symposium on Sediment Microbiology in Cambridge in 1980. The contributors have, in most cases, considerably extended their original verbal contributions and have included recent developments into their manuscripts which were received by the Editors in 1981. The aim of this volume is not so much to provide a comprehensive overview of sediment microbiology but to concentrate on those areas of research which are currently proving fruitful. The multi-disciplinary approach being applied to studies of sediment processes is reflected in a number of the articles. Much of the data presented in this volume raise more questions than answers, but this is to be expected. We hope, however, that the reader will gain some insight into the problems posed, the methods being used to provide solutions, and appreciate the plethora of further work which is required before we can feel that we have at least started to understand the microbiology of sedimentary environments.

D.B. Nedwell and C.M. Brown

1st November 1981

CONTENTS

Chapter 1

THE SEDIMENT ENVIRONMENT

S.J. MALCOLM and S.O. STANLEY

Dunstaffnage Marine Research Laboratory,
Oban, Argyll, Scotland, UK

Introduction

The environment within a sediment is a complex function of many different factors, such as the major mineral matrix, the texture, the amount of organic carbon and the geographical location. In this chapter we briefly outline the nature of the major components of sediments and indicate how these vary between the main environments of recent sediment deposition. In addition, we outline the major diagenetic processes and how these control the chemical environment. It is hoped that the reader will find this introduction to the variety of sediments studied under the heading of sediment microbiology useful when considering the more detailed studies of processes which form the rest of this volume.

Origin of Sediments

A sediment is commonly defined as "solid material that has settled down from a state of suspension in a liquid" [Greensmith, 1971] and, while this definition is limited in the geological usage, it is adequate for most sediments studied by microbiologists.

The geomorphology and climate of the present world, with its young, high mountain chains and active mechanical denudation, leads to the domination of sedimentary environments with sediments formed from detrital minerals. Only in the deep sea and certain nearshore areas, such as coral reefs, does the predominance of detritus give way to sediments of biogenic origin.

Sediments consist of three major components: *detrital* material derived from the erosion of the continents, *biogenic* material that is formed by biological productivity and *authigenic* material that is formed *in situ*. The final character of the sediment, and hence its environment, is related to the relative proportions of these components which, in turn, depend on the environment of deposition.

Detrital material consists mostly of alumino-silicate minerals (Table 1) carried by rivers to lakes and to the continental margins or carried by winds to the deep seas. It is spread to all depositional environments, in greatest amount near continents and less abundant in the deep oceans. The mineralogical composition of sediments therefore tends to reflect the composition of the source rocks because the alumino-silicates are more or less stable under sedimentary conditions. The clay minerals are the most reactive and, because of their layered structure, have a very large surface area. They are ion exchangers and play a role in controlling the distribution of, for instance, the major cations (Na^+, K^+, Ca^{2+}, Mg^{2+}) and ammonium (NH_4^+) and provide adsorption sites for organic molecules [Meyers and Quinn, 1971; Button, 1969], trace metals [e.g. Chester, 1965] and bacteria.

Biogenic components are represented by three important substances that are produced as the hard and soft tissues of free floating plankton in the surface waters of lakes, estuaries and seas. These materials are calcite, opal (cryptocrystalline SiO_2) and organic matter. Calcite is formed by foraminiferids and coccolithophorids; opal is formed by radiolarians and diatoms and organic material is formed by all organisms. The type and amount of organic matter in the sediment is one of the most important factors in control of the sediment environment. Two varieties of organic matter can be considered; that due to primary production and the terrigenous organic detritus which is brought to the site of deposition from the continents. These organic components have a differing structure and composition. For example terrigenous organic matter generally has a high carbon:nitrogen ratio (17-35) compared to the organic matter of primary production which has a low carbon:nitrogen ratio (5-10). These differences may be important when considering the ease of degradation of the organic matter. It is considered that the materials with lower C:N ratios are more readily attacked than the materials with a high C:N ratio.

Authigenic materials consist of mineral phases precipitated either at the sediment water interface or within the body of the sediment. These minerals form a very small but interesting fraction of a sediment. Authigenic minerals commonly found in sediments are listed in Table 1.

Environments of Deposition

The environment of deposition is of great importance in determining the physical character of a sediment, that is its state of division (grain size distribution), its texture and organic content. The chief environments of

TABLE 1

The main detrital aluminosilicate
and common authigenic minerals present in recent sediments

Alumino-silicate minerals		Authigenic minerals	
Quartz	SiO_2	Reddingite	$Mn_3(PO_4)_2.3H_2O$
Feldspars:-		Vivianite	$Fe_3(PO_4)_2.8H_2O$
Orthoclase	$KAlSi_3O_8$	Mackinawite	FeS
Plagioclase	$NaAlSi_3O_8$	Greigite	Fe_3S_4
Clay minerals:- Kaolinite	$Al_2Si_2O_5(OH)_4$	Pyrite	FeS_2
Illite	$KAl_3Si_3O_{10}(OH)_2$	Alabandite	MnS
Montmorillonite	$Al_2Si_4O_{10}(OH)_2.nH_2O$	Hauerite	MnS_2
Chlorite	$Mg_5Al_2Si_3O_{10}(OH)_8$	Siderite	$FeCO_3$
		Rhodochrosite	$MnCO_3$

deposition are shown in Table 2. Each of the environments of deposition has characteristics which contribute to the conditions in the sediments.

TABLE 2

The major environments of recent sediment deposition

Continental	Fluviatile (rivers) Lacustrine (lakes)	Fresh water
Intermediate	Estuarine (estuaries and fjords) Tidal flats	
Marine	Littoral (beaches and intertidal zone) Shallow water (continental shelves) Abyssal (deep sea floor)	Salt water

Continental Environments

The continental environments of deposition are relatively minor in terms of area but are extremely important because of the tremendous industrial and recreational use that is made of them. The sediments of rivers are very inhomogenous and in a state of constant motion. It is therefore difficult to characterize a specific fluviatile sediment environment. The most important features of riverine sediments are that they are detrital, closely resemble the source rocks and contain freshwater.

Lacustrine environments are low energy and generally show a gradation of grain size from the littoral to the profundal zones - the sediments in the deepest parts of the lakes being the finest grained. The amount of organic material in the sediment depends not only on the quantity carried by rivers but also on the nature of the lake. An oligotrophic lake with low primary production will have less organic matter available than a eutrophic lake with high primary production.

Intermediate Environments

Estuaries are very important environments of deposition, for not only are they areas of great industrial and recreational use but they are also the interface between the fresh and salt water regimes and, therefore, the site of important chemical processes. Like rivers, estuaries are very dynamic and this is reflected in the diversity of sediments from gravels through to the finest muds. There is great inhomogeneity both spatially and with time. Estuaries are areas of high biological productivity and this leads to a large amount of organic material entering the sediments.

Tidal flats are an important intermediate environment bordering both estuaries and tidal seas where the land topography is low. Again a great variety of sediment types are present, ranging from shell gravels to muds, fine grained silts being the dominant type present. These sediments are unique in that they are sub-aerially exposed for a large amount of the time and subject to flows of ground water. These sediments usually contain a high proportion of organic matter in the form of debris from neighbouring salt marshes. The influence of primary production is small.

Marine Environments

The marine environment contains by far the greatest area of deposition and this is reflected in the diversity of sediment types. Marine environments may be considered under two main headings; shallow water and abyssal. Shallow water environments include the littoral areas, where gravels and sands of local derivation accumulate, and the large areas of continental shelf below the depth of most wave action. On the continental shelves the currents acting on pre-existing sediments cause a large-scale sorting into areas of gravel, sand and mud. The distribution is determined by topography and current strengths; that is, in areas of high current velocity there are areas of gravel left as lag deposits. The material undergoing sorting on the continental shelves is the debris left from Pleistocene glaciation. The distribution of major sediment types in the North Sea area is shown in Fig. 1. In the tropics there is a tendency to domination by carbonate sediments.

The amount of organic material entering continental shelves varies greatly depending on local hydrography and production and is, to some extent, controlled by the availability of nutrients. Thus, in areas of upwelling such as off south-west Africa, the intense upwelling of nutrient rich water leads to the incorporation of as much as 26.5% (w/w) organic carbon into the sediments [Calvert and Price, 1970].

Away from the continents the domination of the

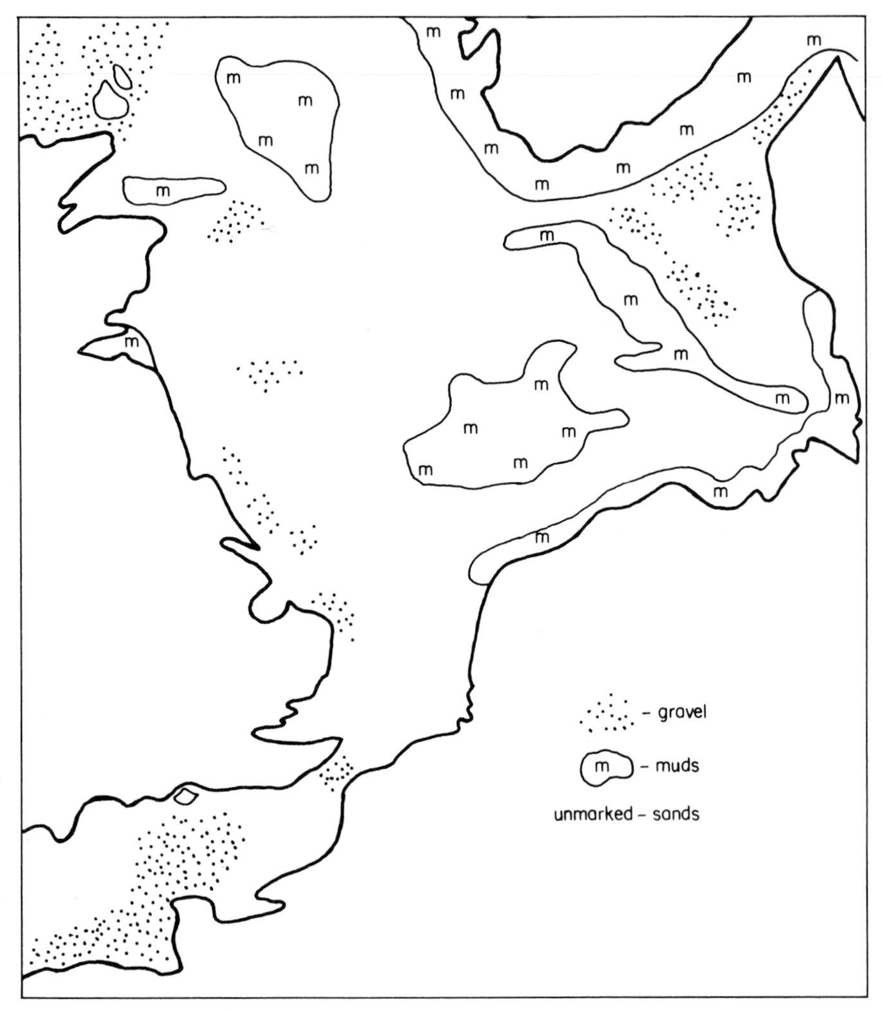

Fig. 1 The distribution of the major sediment types in the North Sea. Redrawn after Greensmith [1971].

sedimentary regime by alumino-silicates is subject to the input of the biogenic components, calcite and opal. Alumino-silicates are transported to the deep sea mostly via the atmosphere and therefore tend to be the fine grained clay minerals. Deposition of this material leads to the formation of "red" clay over the entire sea floor at a rate of about 0.3 cm^{-2} $1000y^{-1}$ [Broecker, 1974]. The type of sediment found in the deep sea however, reflects the overlying primary production. Much of the sea floor is covered by sediments which are characterized as carbonate ooze or siliceous ooze, depending on the dominant biogenic component. The distribution of opal in the deep sea sediments (Fig. 2a) reflects almost directly the distribution in surface waters of the production of siliceous organisms. The distribution of calcite reflects (Fig. 2b) not only the production of calciferous organisms but also the water depth, as the solubility of calcium carbonate is pressure dependant.

The amount of organic matter in deep sea sediments is very small, averaging about 0.3% (w/w) of the sediment. This is due to the long residence time of much of the organic matter in the water column and its degradation by aerobic processes. Water depth plays an important role in determining the organic content of underlying sediments.

Diagenetic Processes

The important chemical reactions which occur during diagenesis of sediments are related to the nature of the sedimentary organic matter and the pathway of its decomposition. A major consequence of the oxidation of organic matter in sediments is the release of the products to the sediment interstitial waters and consumption of various electron accepting species. The interstitial fluids form a significant component of sediments – up to 90% (w/w). The direct effect of these reactions is to promote a flow of dissolved chemicals across the sediment/water interface. The flux is controlled by the availability of organic material, the rate of degradation and by the diffusion/advection properties of the sediment.

Degradation of Organic Matter

The simplest model of diagenetic reactions in sediments is one where organic matter of an assumed (rarely known) composition is oxidized by a sequence of oxidants, which can be predicted based on the free energy yield of the reactions. The reactions producing the greatest amount of energy dominate until that oxidant is consumed and the next most efficient reaction takes over. The reactions are listed in order in Table 3. Particular reactions dominate in specific environments, thus in the marine realm, sulphate is a major oxidant after the consumption of oxygen while in freshwater sediments sulphate reduction is less important due to the much lower ambient concentration of sulphate in freshwater.

The organic matter undergoing degradation in Table 3 is the metabolizable organic matter (MOM), the fraction of the total which can be readily utilized [Berner, 1977]. The MOM may be a relatively small fraction of the total organic content of a sediment. It is difficult to estimate the amount of MOM available in a sediment and also difficult to estimate its composition. However, metabolite concentrations in the interstitial water may be used to make an approximation to the gross composition

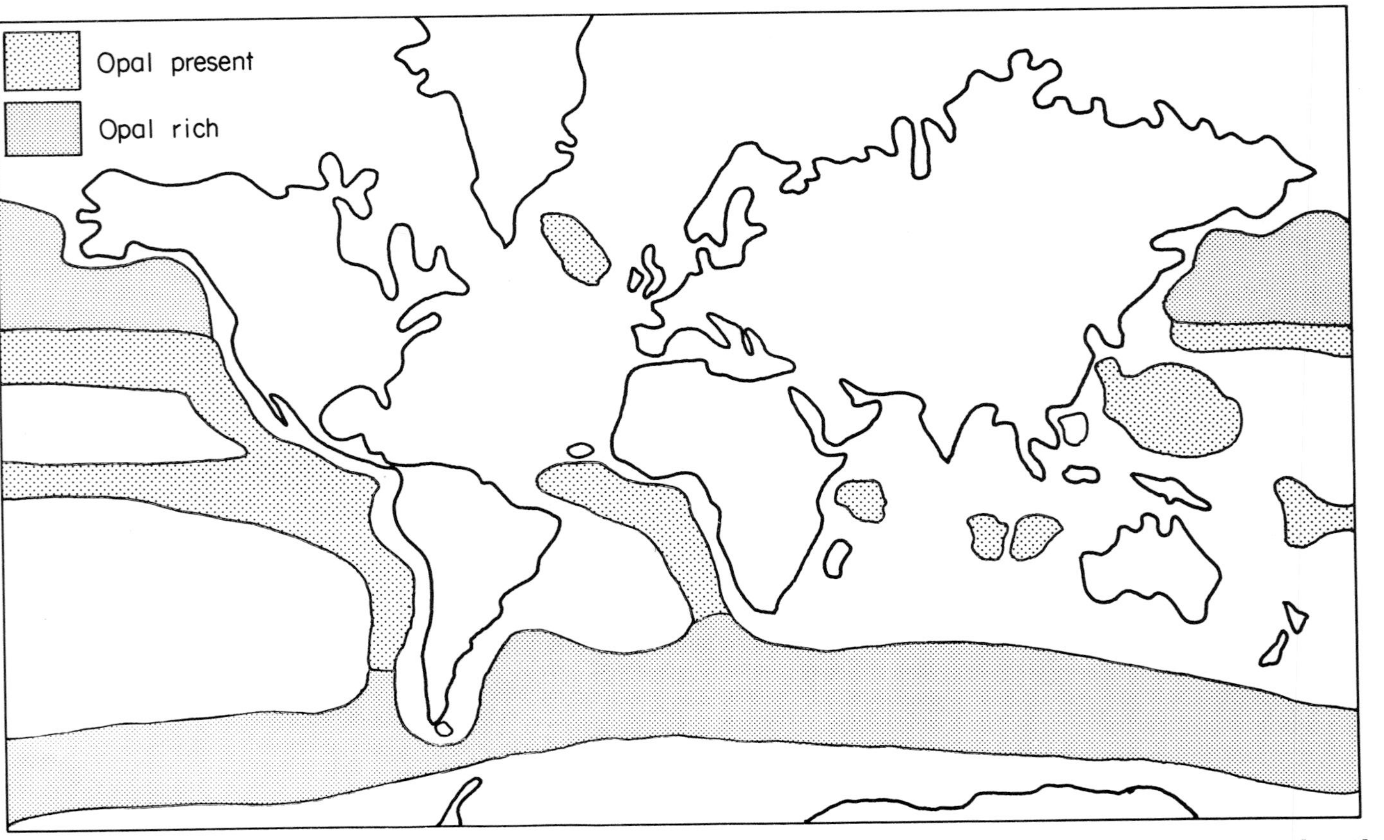

Fig. 2 a) The distribution of biogenic opal in the sediments of the deep sea. Redrawn after Broecker [1974].

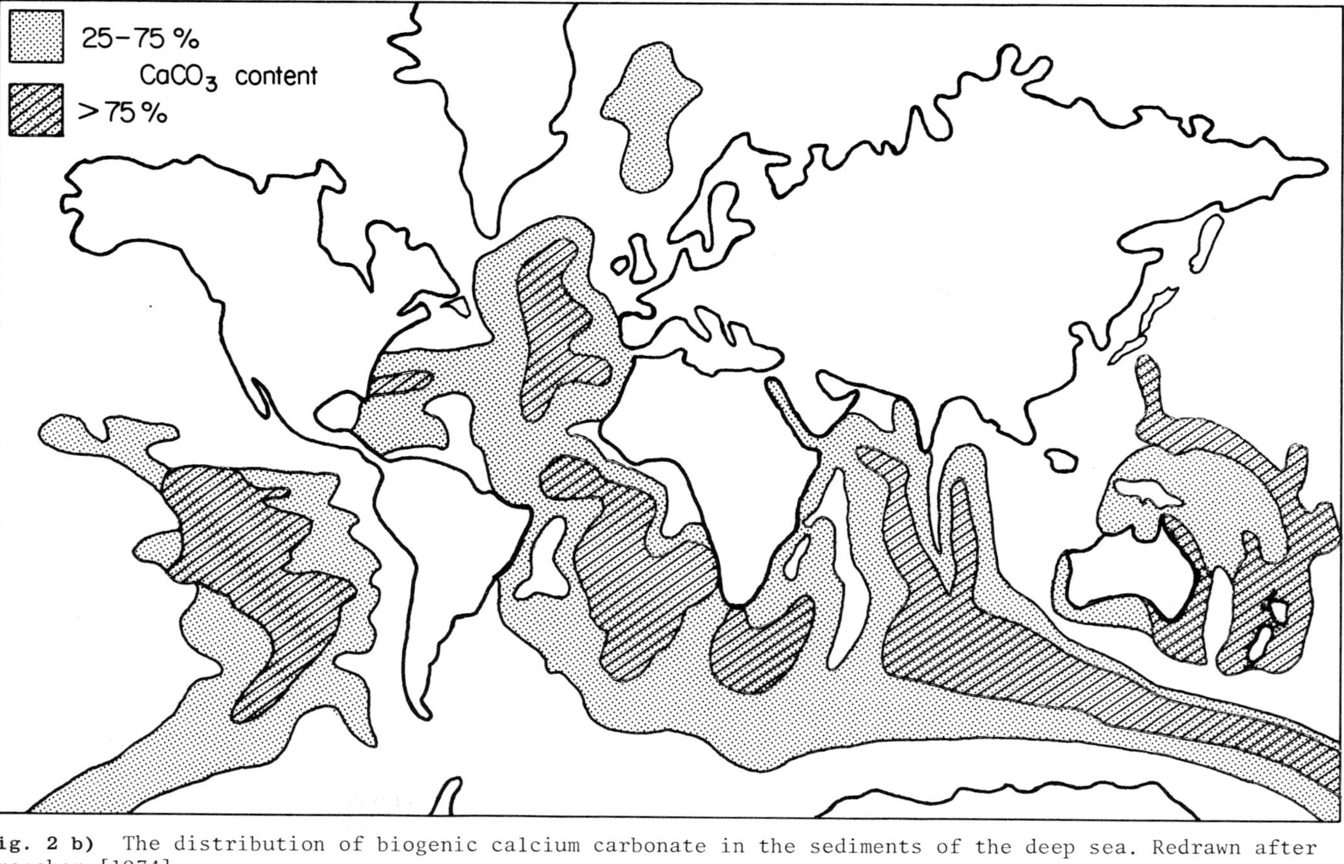

Fig. 2 b) The distribution of biogenic calcium carbonate in the sediments of the deep sea. Redrawn after Broecker [1974].

TABLE 3

The major model disgenetic reactions for the degradation of organic matter in recent sediments. The composition of metabolizable organic matter shown is the composition of living marine plankton. [after Redfield, 1958]

Zone	Reactants	Products
Aerobic zone	$(CH_2O)_{106}(NH_3)_{16}H_3PO_4 + 106\ O_2$	$106\ CO_2 + 16\ NH_3 + H_3PO_4 + 106\ H_2O$
	$(CH_4 + 2\ O_2$	$CO_2 + 2\ H_2O)$
Manganese reduction	$(CH_2O)_{106}(NH_3)_{16}H_3PO_4 + 212\ MnO_2 + 332\ CO_2 + 120\ H_2O$	$438\ HCO_3^- + 16NH_4^+ + HPO_4^{2-} + 212\ Mn_4^{2+}$
Nitrate reduction zone	$(CH_2O)_{106}(NH_3)_{16}H_3PO_4 + 84.8NO_3^-$	$7.2\ CO_2 + 98.8\ HCO_3^- + 16NH_4^+ + 42.4\ N_2 + HPO_4^{2-} + 49\ H_2O$
Iron reduction	$(CH_2O)_{106}(NH_3)_{16}H_3PO_4 + 424Fe(OH)_3 + 756\ CO_2$	$862\ HCO_3^- + 16\ NH_4^+ + HPO_4^{2-} + 424\ Fe^{2+} + 304\ H_2O$
Sulphate reduction zone	$(CH_2O)_{106}(NH_3)_{16}H_3PO_4 + 53\ SO_4^{2-}$	$39\ CO_2 + 67\ HCO_3^- + 16\ NH_4^+ + HPO_4^{2-} + 53\ HS^- + 39\ H_2O$
	$(CH_4 + SO_4^{2-}$	$HCO_3^- + HS^- + H_2O)$
Carbonate reduction zone	$(CH_2O)_{106}(NH_3)_{16}H_3PO_4 + 14\ H_2O$	$39\ CO_2 + 14\ HCO_3^- + 53CH_4 + 16NH_4^+ + HPO_4^{2-}$
(Methane fermentation)	$CO_2 + 4\ H_2$	$CH_4 + 2\ H_2O$

of the MOM. In the case of no sediment adsorption, the composition of MOM can be deduced from the diffusion corrected regeneration stoichiometry. For example, if the rates of the fluxes of total carbon dioxide (ΣCO_2) and ammonia (NH_3) equal the relative regeneration rates [Berner, 1977], then the carbon:nitrogen ratio of the degrading organic material is given by

$$\frac{C}{N} = \frac{F_{\Sigma CO_2}}{F_{NH_3}} = \frac{D_{\Sigma CO_2}}{D_{NH_3}} \times \frac{(dC/dz)_{\Sigma CO_2}}{(dC/dz)_{NH_3}} = \frac{D_{\Sigma CO_2}}{D_{NH_3}} \times \frac{dC_{\Sigma CO_2}}{dC_{NH_3}}$$

and can be estimated from concentration gradients and diffusion coefficients. This will only be true if the sediment is diffusion limited.

The models shown in Table 3 allow us to define reaction zones, (environmental zones) in the sediments and the models are an aid in determining the position of important boundaries. For example, the nitrate concentration in interstitial water increases to a maximum at the base of the zone of aerobic respiration and decreases in the zone of denitrification [Bender *et al.*, 1977]. The nitrate maximum will thus mark the boundary between oxic and anoxic sediment.

Transport in Sediments

The texture of a sediment, a feature consisting of such properties as grain size distribution, grain shape and sediment packing, affects the diffusion of solutes in the sediment interstitial water by determining the sediment porosity, permeability and tortuosity. The open-ness of a sediment to input or output is therefore controlled by its texture. Adsorption also affects the movement of solutes in a sediment; the adsorption capacity of a sediment is largely determined by the amount of clay mineral it contains and, hence, may also be related to sediment texture. When considering diffusive processes in sediments the Fickian diffusion coefficient has to be modified to take account of adsorption [Berner, 1976] and tortuosity [Goldhaber *et al.*, 1977] (texture) to give a "bulk sediment diffusion coefficient".

In many sediments, particularly in shallow water, there is an upper zone exhibiting depth independent interstitial water profiles to depths as great as 30 cm [Goldhaber *et al.*, 1977; Malcolm, 1981] which are not amenable to interpretation by the simple diffusion/ reaction models. These profiles arise from irrigation and/or particle mixing by macro-infaunal organisms; the process termed bioturbation [Goldhaber *et al.*, 1977; Aller, 1978]. As an example, consider the profiles of interstitial sulphate (SO_4^{2-}) shown in Fig. 3. The upper

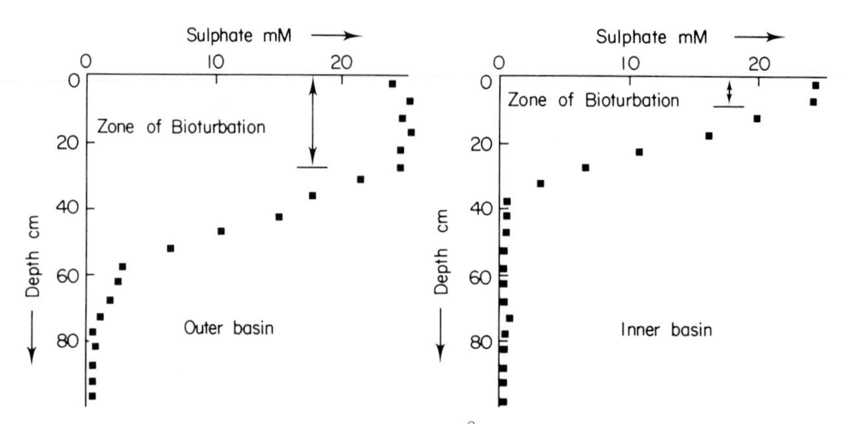

Fig. 3 Depth profiles of dissolved SO_4^{2-} in sediments from the inner and outer basins of Loch Etive, a fjordic estuary on the west coast of Scotland. Both profiles show depth independent upper zones due to macrofaunal irrigation of the sediments.

zone of each sediment has been shown by sediment incubation experiments to support rapid sulphate reduction, but is subject to intense irrigation by populations of tube dwelling polychaetes. In these sediments the rate of irrigation of the sediment is much greater than the apparent rate of sulphate reduction as the concentration of dissolved sulphate is maintained close to the concentration found in the overlying water. The depths of the upper zone in the two sediments are probably controlled by differences in the faunal assemblage [Malcolm, 1981]. The rate of interstitial transport may be measured using a ^{22}Na tracer technique and rates of "biopumping" of up to 0.7 $cm^3 cm^{-2} day^{-1}$ have been reported [McCaffrey *et al.*, 1980]. The most reactive fraction of organic matter is generally consumed in the zone of bioturbation [Malcolm, 1981], and greater than 90% of the sediment metabolic activity occurs in the upper zone [McCaffrey *et al.*, op. cit.].

Sediments vary from completely oxic, where the supply of organic carbon is less than the supply of oxygen, as in deep-sea sediments, to wholly anoxic, where the supply of organic carbon is much greater than the supply of oxygen, as in near shore sediments where anaerobic processes are often dominant. In the basins of restricted circulation, that is some lakes and fjord estuaries, basins on continental shelves (e.g. the Cariaco Trench) and the Black Sea, the oxic/anoxic boundary moves upward out of the sediment into the water column. The sediments of these basins are devoid of macrobenthic life and provide the rare occurrence of sediment environments that are controlled simply by diffusive processes.

Effect of Organic Degradation

The changing chemistry of the sediments is reflected in changes in the "redox potential" and the pH. The dominant pH controlling species in many sediments are carbon dioxide (CO_2) and ammonia (NH_3), the concentrations of which may vary with the nature and pathway of degradation of the organic matter. The pH of sediments is generally lower than the pH of the overlying water, if there is active organic decay. The "redox potential" is determined by various electrochemical couples formed from a small group of dissolved species (e.g. HCO_3^-, CH_4, HS^-, H_2S, NO_3^-, NH_4^+, O_2). The "redox potential" (Eh) generally decreases in sediments where organic matter is being destroyed. The measurement of Eh is however very difficult, due to the inhomogeneity of sediments and the development of microenvironments of very different potential to those of the bulk sediment. Bulk sediment "Eh" measurements are useful in giving an indication of the amount of degrading organic matter present and have been successfully used as an indicator of the extent of organic pollution in near-shore marine sediments [Pearson and Stanley, 1979].

The concentration of many species (HCO_3^-, HS^-, HPO_4^{3-}, $HSiO^-$, NH_4^+, NO_3^-, Fe^{2+}, Mn^{2+}, dissolved organic compounds, trace metals) is often much higher in the sediment interstitial water than in the overlying water and this leads to major fluxes across the sediment water interface. In nearshore areas this regeneration of nutrient elements from sediments may be an important factor controlling the productivity of the overlying water. The high concentrations of solutes also leads to supersaturation with respect to certain mineral phases which precipitate forming the authigenic fraction of the sediments. The interstitial water is a relatively enriched medium into which most of the chemically labile and biologically important compounds and elements are mobilized during diagenesis and may thus be an important control on the dominant bacterial activities present in the sediments.

The sediment environment is a very complex mixture of features which are, in themselves, highly variable - it might be observed that there is more variability in sedimentary systems than there is uniformity. Thus the study of sediment microbiology is burdened by the need to know the sedimentology and geochemistry of the deposits.

References

Aller, R.C. (1978). Experimental studies of changes produced by deposit feeders on pore water, sediment, and overlying water chemistry. *American Journal of Science* **278**, 1185-1234.

Bender, M.L., Fanning, K.A., Froelich, P.N., Heath, G.R. and Maynard, V. (1977). Interstitial nitrate profiles and oxidation of sedimentary organic matter in the eastern equatorial Atlantic. *Science* **198**, 605-609.

Berner, R.A. (1976). Inclusion of adsorption in the modelling of early diagenesis. *Earth and Planetary Science Letters* **29**, 333-340.

Berner, R.A. (1977). Stoichiometric models for nutrient regeneration in anoxic sediment. *Limnology and Oceanography* **22**, 781-786.

Broecker, W.S. (1974). "Chemical Oceanography" Harcourt, Brace, Jovanovich, Inc. New York.

Button, D.K. (1969). Effect of clay on the availability of dilute organic nutrients to steady state heterotrophic populations. *Limnology and Oceanography* **14**, 95-100.

Calvert, S.E. and Price, N.B. (1970). Recent sediments of the south-west African shelf. In "The Geology of the East Atlantic Continental Margin" (Ed. F.M. Delany), Vol.4, Africa, pp.173-185. Institute of Geological Sciences Report No. 70/16.

Chester, R. (1965). Adsorption of zinc and cobalt on illite in sea-water. *Nature* **206**, 884-886.

Goldhaber, M.B., Aller, R.C., Cochran, J.K., Martens, C.S. and Berner, R.A. (1977). Sulphate reduction, diffusion and bioturbation in Long Island Sound sediments: Report of the FOAM group. *American Journal of Science* **277**, 193-237.

Greensmith, V.T. (1971). "Petrology of the Sedimentary Rocks" p.502. George Allen and Unwin Ltd.

Malcolm, S.J. (1981). The chemistry of the sediments of Loch Etive, Scotland. p.183. Unpublished Ph.D. Thesis, University of Edinburgh.

McCaffrey, R.J., Myers, A.C., Davey, E., Morrisson, G., Bender, M.L., Luedtke, N., Cullen, D., Froelich, P.N. and Klinkhammer, G. (1980). Benthic fluxes of nutrients and manganese in Narragansett Bay, Rhode Island. *Limnology and Oceanography* **25**, 31-44.

Meyers, P. and Quinn, J.G. (1971). Fatty acid-clay mineral association in artificial and natural seawater solutions. *Geochemica et Cosmochimica Acta* **35**, 628-632.

Pearson, T.H. and Stanley, S.O. (1979). Comparative measurement of the redox potential of marine sediments as a rapid means of assessing the effect of organic pollution. *Marine Biology* **53**, 371-379.

Chapter 2

MODELLING THE PROCESSES OF ORGANIC MATTER DEGRADATION AND NUTRIENTS RECYCLING IN SEDIMENTARY SYSTEMS

G. BILLEN*

Laboratory of Oceanography, University of Brussels, 50, Avenue F.D. Roosevelt, 1050 Brussels, Belgium

Introduction

Evaluating the rates of microbial processes in biogeochemical cycles and understanding the factors controlling them, are often the final aims of ecological studies of sediment microbiology. This information is required for a full description of aquatic ecosystems and for predicting the changes they can undergo under natural or man-induced stresses.

This kind of information (rates of processes, fluxes, etc.) is dynamical by nature and is difficult to measure in *in situ* conditions. The development of convenient techniques for determining *in situ* rates of microbial activity can be considered as the limiting step in the progress of our knowledge in this field. Existing methods involve a) the *in situ* (or *quasi in situ*) determination of microbial processes at particular depth of a sediment core [Billen, 1976; Wieser and Zech, 1976; Sørensen, 1978; Jørgensen, 1978; Christensen and Blackburn, 1980] b) the direct measurement of fluxes across the sediment water interface by means of sediment traps for evaluating the flux of particulate organic material to the bottom [for instance Bloesch, 1978; Hargrave and Burns, 1979...] c) the use of belljars and other devices for evaluating either the fluxes of dissolved nutrients to the water column [Carey, 1967; Fanning and Pilson, 1974; James, 1974; Rowe *et al.*, 1975; Seitzinger *et al.*, 1980...] or the benthic consumption of oxygen [Smith *et al.*, 1972; Hargrave, 1973]. These methods however, are often tedious, and the simultaneous determination on a single sediment core of all rates necessary to make up a balance of the transformation even of a single element, is impossible in most cases.

---

* Chercheur qualifié of the Fonds National de la Recherche Scientifique.

On the other hand, statistical data such as the concentration of substances, either in the particulate phase or in the interstitial water of sediments, is far more easy to measure accurately. It is therefore tempting to use such data for deducing the required dynamical information. This can be done by analysis of the vertical distribution of substances within the sedimentary column, provided that at least one of the dynamic parameters involved is known.

Stratigraphical analysis is an example of this approach. The distribution of sediment properties with depth is regarded as reflecting their evolution with time. This evolution can be quantified if the conditions of sedimentation are reasonably constant and if the deposition rate (ω) is known. Rittenberg *et al.* [1955] have used this approach for evaluating organic nitrogen recycling in Santa Barbara basin sediments. However, the time scale of most microbial processes involved in biogeochemical cycles in sediments is much shorter than the usual deposition rates and most of the microbiological activity in sediments occurs in the top 5 to 10 cm layer. Moreover, this upper layer is submitted to reworking processes (resuspension and bioturbation) which disturb stratigraphy. This is also the case for substances dissolved in the interstitial water, which undergo still more important dispersion processes. A complete analysis of concentration data is therefore necessary, taking into account the distribution of a substance with depth as the result of the processes of production, consumption, vertical dispersion and burial. This is the principle of mathematical diagenetic modelling.

The purpose of this chapter is to present the application of simple mathematical diagenetic analysis as a possible approach to the study of microbial ecology of sediments.

Mixing Processes in Sediments

Mathematical Formalism

A complete and rigorous development of the mathematical theory of diagenesis is far beyond the scope of this chapter. It has been the subject of a number of specialized papers [Berner, 1964, 1971, 1974, 1975; Anikouchine, 1967; Tzur, 1971; Imboden, 1975; Lerman, 1979] and a textbook [Berner, 1980*a*].

The basic procedure in modelling the vertical distribution of a given substance either dissolved in the pore water or included in the solid phase of the sediment is to express the mass balance of that substance at each depth, as a result of reaction, advection and mixing.

Although other approaches have been used by certain

authors* it is customary to consider the mixing processes of both the solid particles and the pore water of the sediment as random and isotropic processes. With this assumption, the formalism of Fick's law can be applied, and the vertical mixing flux J(z) at any depth z' expressed as proportional to the concentration gradient:

$$J(z') = -D \left[\frac{\delta C}{\delta z}\right]_{z'}$$

where C is the concentration of the substance considered, expressed in mass per unit sediment volume, and D is an apparent mixing coefficient, the meaning and magnitude of which is discussed below.

With this formalism for mixing processes, the expression of the mass balance in a reference system linked to the water-sediment interface can be written:

$$\frac{\delta C}{\delta t} = \frac{\delta}{\delta z}\left[D \frac{\delta C}{\delta z}\right] - \frac{\delta}{\delta z}[vC] + R(z) \qquad (1)$$

where t is the time,
R(z) is the resultant rate of all production and consumption processes involving the substance considered,
v is the rate of flow relative to the water-sediment interface.

When compaction of the sediment with depth is neglected (which is quite correct for sands and a first approximation for muds in the upper 10 to 20 cm layer), v is equal to the rate of deposition, ω, which is independent of depth.

If it is assumed that a stationary state has been reached by the concentration profile C(z), this stationary state is the solution of the equation:

$$0 = \frac{d}{dz}\left[D\frac{dC}{dz}\right] - \omega \frac{dC}{dz} + R(z) \qquad (2)$$

For this assumption to be true, the parameters D, ω, and R, and the limit conditions of the equation have to be time independent, a situation which seldom occurs. However, this assumption can be considered as a good approximation in two situations which can be characterized by comparing the turnover time of the considered substance in the sediment (ratio of the standing mass of the substance to its total rate of production or consumption) with the characteristic time of the variations of biological, chemical or physico-chemical conditions:

---

* See for instance the cylindrical-burrow irrigation model of Aller [1978] and the biopumping model of McCaffrey *et al.* [1980].

(i) when the turnover time is great compared with the period of variation of the environmental conditions. In this case, the concentration profile of the substance in a sense "integrates" these variations, and remains very close to the stationary state corresponding to the mean value of the environmental parameters.
(ii) when the turnover time of the considered species is small with respect to the period of variation of the environmental conditions. In this case, the profile rapidly fits to the external conditions and is always very close to the stationary state with respect to instantaneous conditions.

When the analytical expression C(z) of the concentration profiles and all the parameters involved are known for the substance considered, it is possible to calculate its flux across the sediment-water interface by the relation:

$$J\ (0) = -\ D\ \left[\frac{dC}{dz}\right]_{z=0} + \omega C \qquad (3)$$

The integrated value of the biological or chemical production rate of the considered species on the whole sediment column can also be calculated by the relation:

$$I = \int_0^\infty R(z)dz \qquad (4)$$

This allows an overall balance of the transformations and transfers of the substance to be computed.
The quantitative analysis of the information contained in vertical concentration profiles in sediments requires a good knowledge of mixing and dispersion processes affecting both the solid phase and the interstitial phase of the sediments.

Mixing of the Solid Phase

Two kinds of processes can cause the mixing of solid particles: hydrodynamical phenomena inducing resuspension, and biological activities of invertebrates (bioturbation).
Resuspension of sediments occurs when the shear-stress at the water-sediment interface is high enough to overcome the inertia and cohesion of the solid particles [Bagnold, 1966; Allen, 1970]. A critical value of the shear-stress above which resuspension occurs can be defined, depending on the granulometry of the sediment and its cohesion. The shear-stress at the water-sediment interface on the other hand depends on the current velocities in the water column, influenced by meteorological conditions. In shallow areas, the critical

shear-stress can be reached at each storm. In the case of fine muds, characterized by strong cohesion, shear-stress at the bottom can result in the permanent occurrence of a perturbated layer of a few cm depth, with higher porosity [Vanderborght *et al.*, 1977].

Important reworking of the upper layer of sediments results from burrowing and feeding activities of benthic invertebrates. Many authors have estimated the amount of sediment worked per unit area and unit time by a given population of benthic organisms. From this, a reworking- or turnover-time of the sediment can be calculated, taking into account the depth accessible to the organisms. The last is generally limited to 5 to 15 cm [Dapples, 1942; Arrhenius, 1963; Laughton, 1963] although organisms can occasionally penetrate down to 30 cm or more, especially in sands. Turnover times determined in various coastal marine environments are in the range 10 weeks to 15 years [Nichols, 1974; Rhoads, 1963, 1967, 1973; Gordon, 1966]. By dimensional analysis of such data, Guinasso and Schink [1975] and Aller and Cochran [1976] estimated Fickian-like mixing coefficients for near-shore marine sediments in the range 3×10^{-7} to 5×10^{-5} cm^2 sec^{-1} (see Table 1a). By modelling the distribution of some tracers with depth in sediments, either in environmental situations or in laboratory experiments, other authors have determined mixing coefficient of the solid phase by adjustment to the experimental data [Duursma and Gross, 1971; Luedtke and Bender, 1979; Aller *et al.*, 1980]. Their estimations, gathered in Table 1a, are in the range 3×10^{-7} to 2×10^{-6} cm^2 sec^{-1}. Table 1a shows also some values of mixing coefficient obtained for deep-sea sediments, which are two or three order of magnitude lower than those for near shore-sediments.

Both processes described above are restricted to the upper layer of the sediments. Data of Table 1a show that this layer is about 2 to 5 cm deep for muds and 30 to 40 cm for sands. No mixing of the solid phase occurs at deeper layers. In most diagenetic models concerned with the whole sedimentary column, the sediment must be considered as made of two layers: a perturbated upper layer of a few to a few tens cm depth where mixing of the solid phase occurs, underlain by a deeper unmixed layer [Goldberg and Koide, 1962; Berger and Heath, 1968; Guinasso and Schink, 1975; Nozaki *et al.*, 1977].

Mixing Processes in the Interstitial Phase

Some mixing within the interstitial water of sediments always occurs through molecular diffusion, but can be greatly enhanced either through purely hydrodynamical processes or through benthic biological activities.

The molecular diffusion of most ions or dissolved species in the pore water of sediments can be

TABLE 1

Values of apparent vertical dispersion coefficient for (a) the solid phase, D_s, (b) the interstitial phase, D_i, as determined in the upper layer of natural sediments by various authors

Location	Sediment type	Depth (cm)	D (cm^2 sec^{-1})	Authors
(a) Solid particles mixing (D_s)				
Near shore sediments				
Chesapeake Bay	-	-	10^{-6}	Duursma and Gross [1971]
Buzzards Bay	-	0 - 2	$3x10^{-8}$	Guinasso and Schink [1975]
Long Island Sound	mud	0 - 2	$2x10^{-7}$	Guinasso and Schink [1975]
Barnstable Harbor	-	0 - 6	$0.8x10^{-7}$	Guinasso and Schink [1975]
Holy Island Sands	sand	0 - 38	$4x10^{-5}$	Guinasso and Schink [1975]
Caves Haven	sand	0 - 38	$0.7x10^{-5}$	Guinasso and Schink [1975]
Fresh water Lake	mud	0 - 6	$4.4x10^{-5}$	Guinasso and Schink [1975]
Narragansett Bay	mud	0 - 5	$3x10^{-7}$	Luedtke and Bender [1979]
Long Island Sound	mud	0 - 4	$1.2 - 3.5x10^{-6}$	Aller and Cochran [1976]
Long Island Sound	mud	0 - 5	$1-3x10^{-6}$	Aller *et al.* [1980]
Long Island Sound	mud	0 - 4	$1-3x10^{-7}$	Krishnaswami *et al.* [1980]
Lake Huron	mud	0 - 3/6	$1-2x10^{-7}$	Robbins *et al.* [1977]
Coastal North Sea	sand	0 - 15	$1.4x10^{-7}$	Billen (in press)

Deep sea sediments				
Mediterranean	-	0 - 12	1.2×10^{-8}	Guinasso and Schink [1975]
Atlantic	-		$3-8 \times 10^{-9}$	Guinasso and Schink [1975]
Caribbean Sea	-	0 - 20	$2\ 10^{-11} - 6 \times 10^{-9}$	Guinasso and Schink [1975]
Mid-Atlantic	calc. ooze	0 - 8	6×10^{-9}	Nozaki *et al.* [1977]
Equatorial Pacific	clay-siliceous ooze	-	$8-14 \times 10^{-9}$	Turekian *et al.* [1978]
Antarctic	siliceous ooze	-	$1-8 \times 10^{-9}$	Turekian *et al.* [1978]
(b) Pore water mixing (D_i)				
Long Island Sound	mud	0 - 8	$>2.8 \times 10^{-5}$	Goldhaber *et al.* [1977]
Coastal North Sea	mud	0 - 3.5	10^{-4}	Vanderborght *et al.* [1977]
Coastal North Sea	sand	0 - >15	$0.5 - 2 \times 10^{-4}$	Billen [1978]
Narragansett Bay	mud	0 - 25	4×10^{-5}	McCaffrey *et al.* [1980]
Laboratory experiments				
	silt clay (with added *Yoldia*)	0 - 4	10^{-5}	Aller [1978]
	(with added *Clymenella*)	0 - >11	$2 - 3 \times 10^{-4}$	Aller [1978]

characterized by coefficients in the range 1 to 10 x 10^{-6} cm^2 sec^{-1}, taking into account the tortuosity of the interstitial space [Li and Gregory, 1974; McDuff and Ellis, 1979; Krom and Berner, 1980].

In permeable beds, such as sandy sediments, pressure waves generated in the overlying water can propagate themthemselves and induce percolation of water through the interstitial space, which causes more rapid dispersion than molecular diffusion alone. This is shown by the very simple experiment described in Fig. 1, where the mixing coefficient in the pore water of a sand was measured at increasing agitation rates of the overlying water. Mixing coefficients one order of magnitude higher

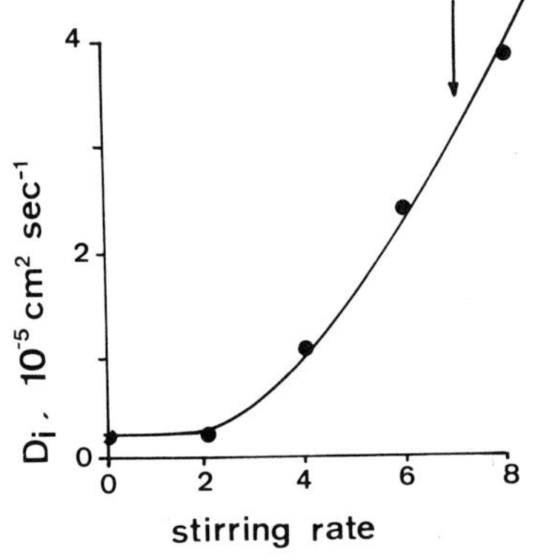

Fig. 1 Apparent mixing coefficient of the interstitial water of a sandy sediment for increasing stirring rate of the overlying water. The arrow indicates the beginning of resuspension of the sediment.

than simple molecular diffusion can be observed at agitation rates insufficient for inducing resuspension of the sediment. Wave induced percolation of water through permeable beds has been demonstrated by *in situ* measurements by Webb and Theodor [1968, 1972], Steele *et al.* [1970], Riedl *et al.* [1972]. It has been theoretically studied by some authors, mostly concerned with the process of wave damping over a permeable bed [Putnam, 1949; Reid and Kajiura, 1957; Riedl *et al.*, 1972]. These pressure fluctuations decrease with depth (z) in the permeable sediment according to a law in exp $(-2\pi z/\lambda)$ where λ is the wave length of the surface waves [Riedl *et al.*, 1972]. They therefore propagate quite deeply into permeable sediments. Accordingly, direct measurements of pressure fluctuations at different depths within a sandy sediment did not show any significant differences between those at the sand surface and those at 20 cm depth [Steele *et al.*, 1970]. This is not the case with impermeable muddy sediments, where these processes are probably restricted to the perturbated layer.

Even in muddy sediments, however, percolation of water

can occur down to a certain depth owing to the pumping activity of invertebrates, particularly polychaetes [Dales, 1961; Mangum, 1964]. This process is called irrigation. The flux of water pumped by organisms across the sediment-water interface can be of several $cm^3\ cm^{-2}\ day^{-1}$ [Luedtke and Bender, 1979; Mangum, 1964].

Table 1b shows the values of Fickian-like mixing coefficients in the interstitial space of sediments resulting from the processes just discussed, as estimated by various authors in coastal sediments.

Organic Matter Degradation

Fate of Organic Matter in Sediments

Organic matter is primarily supplied to the sediments in a particulate form. It is essentially made of macromolecules such as proteins, carbohydrates, fats and nucleic acids. Such compounds cannot be directly taken up by microorganisms, and have first to be hydrolysed into smaller units such as amino-acids, sugars, fatty acids, purine and pyrimidine bases, probably mainly through the action of exoenzymes. Once produced, these low molecular weight, directly usable compounds are rapidly taken up and metabolized by heterotrophic bacteria so that the rate of the overall bacterial process of organic matter degradation is probably controlled by the first limiting step of exoenzymatic hydrolysis. Bacterial biomass can adapt itself very rapidly to the flux of direct substrates produced, maintaining them at low constant concentrations, according to the same mechanisms as those described by Billen *et al.* [1980] for planktonic ecosystems.

Very little is known about the process of exoenzymatic hydrolysis of organic matter in sediments. Its rate is certainly dependent both on the exact nature of the macromolecules present and on their interaction with mineral constituents of the sediments. Adsorption by clay minerals [Linch and Cotnoir, 1956] or complexation by metal ions [Degens and Mopper, 1975, 1976] have been described as processes by which proteins could be protected from enzymatic attack.

Besides being taken up by microorganisms, hydrolytic products of biopolymers can also interact and give rise to complex compounds termed geopolymers (humic acids, fulvic acids, humins and ulmins) [Welts, 1973] which are only slowly biodegraded.

First Order Model of Organic Matter Degradation

Due to poor knowledge of the mechanisms of the first, limiting steps of organic matter degradation in natural environments, it is customary to assume that the rate of the overall process (R) is first order with respect to the biodegradable part of the organic matter (G):

$$R = k\,G \tag{5}$$

This approach has been used widely for modelling organic load degradation in polluted rivers [Streeter and Phelps, 1925] and by most workers on organic degradation in sediments.

In order to take into account the different susceptibilities to bacterial attack of different classes of compounds making up the overall organic matter, Jørgensen [1978*b*] and Berner [1980*b*] have suggested the use of "multi G's - first order kinetics":

$$R = \sum_i k_i G_i \qquad (6)$$

where G_i refers to a particular class of organic compounds, and
k_i is the kinetic constant of degradation of these compounds by bacterial metabolic processes.

The total decomposable organic matter in the sediment is:

$$G = \sum_i G_i$$

The point is that, first in the water column during sedimentation, then in the sediments during burial, the various classes of organic matter are successively exhausted in the order of decreasing biodegradabilities. Table 2 shows some typical values obtained for the rate constants of recently deposited organic matter degradation in the uppermost layer of sediments. These values of k correspond to a class of planktonically derived, easily biodegradable organic material. They all are in the range 0.5 to 2 $year^{-1}$. Measurement of microbial degradation of suspended organic material [Saunders, 1972; Otsuki and Hanya, 1972; Iturriaga, 1979] shows much more important relative rates of 10 to 100 $year^{-1}$.

TABLE 2

Rate constant for recently deposited organic matter degradation in sediments

Location	Sediment type	Depth (cm)	Rate constant (sec^{-1})	Rate constant ($year^{-1}$)	Authors
Long Island Sound	mud	0 - 5	1.4×10^{-8}	0.45	Turekian *et al.* [1980]
Long Island Sound	mud	0 - 5	1.9×10^{-8}	0.59	Berner [1980]
Coastal North Sea	sand	0 - 10	$1.7\text{-}6 \times 10^{-8}$	0.5-2	Billen (in press)

On the other hand, the rate constant of organic matter degradation in the deeper layer of the sediments is several order of magnitude lower than in the upper layer: reported values are in the range 10^{-1} to 10^{-7} $year^{-1}$. Toth and Lerman [1977] and Berner [1978] have shown that a relation exists between the value of k below the bioturbation layer and the deposition rate ω. They explain this by the fact that higher sedimentation rates result in a more rapid burial of readily metabolizable organic material below the bioturbation layer, thus supplying the lower layers with more reactive organic compounds.

The vertical distribution of the various classes of organic matter (G_i) in sediments can be deduced from the diagenetic equation:

$$\frac{\delta G_i}{\delta t} = D_s \frac{\delta G_i^2}{\delta z^2} - \omega \frac{\delta G_i}{\delta z} - k_i G_i \quad \text{for } z \leqslant z_B \tag{7}$$

and

$$\frac{\delta G_i}{\delta t} = - \omega \frac{\delta G_i}{\delta z} - k_i G_i \quad \text{for } z > z_B \tag{8}$$

where D_s is the mixing coefficient of the solid phase in the perturbated layer and z_B, the depth of the perturbated layer. The stationary solution of these equations, with the following boundary conditions:

$$G_i \ (z = 0) = G_{io}$$

$$G_i \ (z = \infty) \text{ remains finite}$$

and imposing continuity at $z = z_B$, is:

$$G_i = G_{io} \exp \left[\frac{\omega - \sqrt{\omega^2 + 4D_s k_i}}{2D}\right] z \quad \text{for } z \leqslant z_B \tag{9}$$

$$G_i = G_i(z_B) \exp \left[- \frac{k_i}{\omega} (z - z_B)\right] \quad \text{for } z > z_B \tag{10}$$

if $\omega << 2\sqrt{k_i D_s}$, the expression (9) of G_i in the perturbated layer reduces to:

$$G_i = G_{io} \exp \left[- \sqrt{\frac{k_i}{D_s}} z\right] \quad \text{for } z \leqslant z_B \tag{11}$$

The vertical profile of total organic carbon concentration $G = \Sigma_i G_i$ displays therefore a succession of slower and slower exponential decreases.

A typical situation is that shown in Fig. 2 for Long

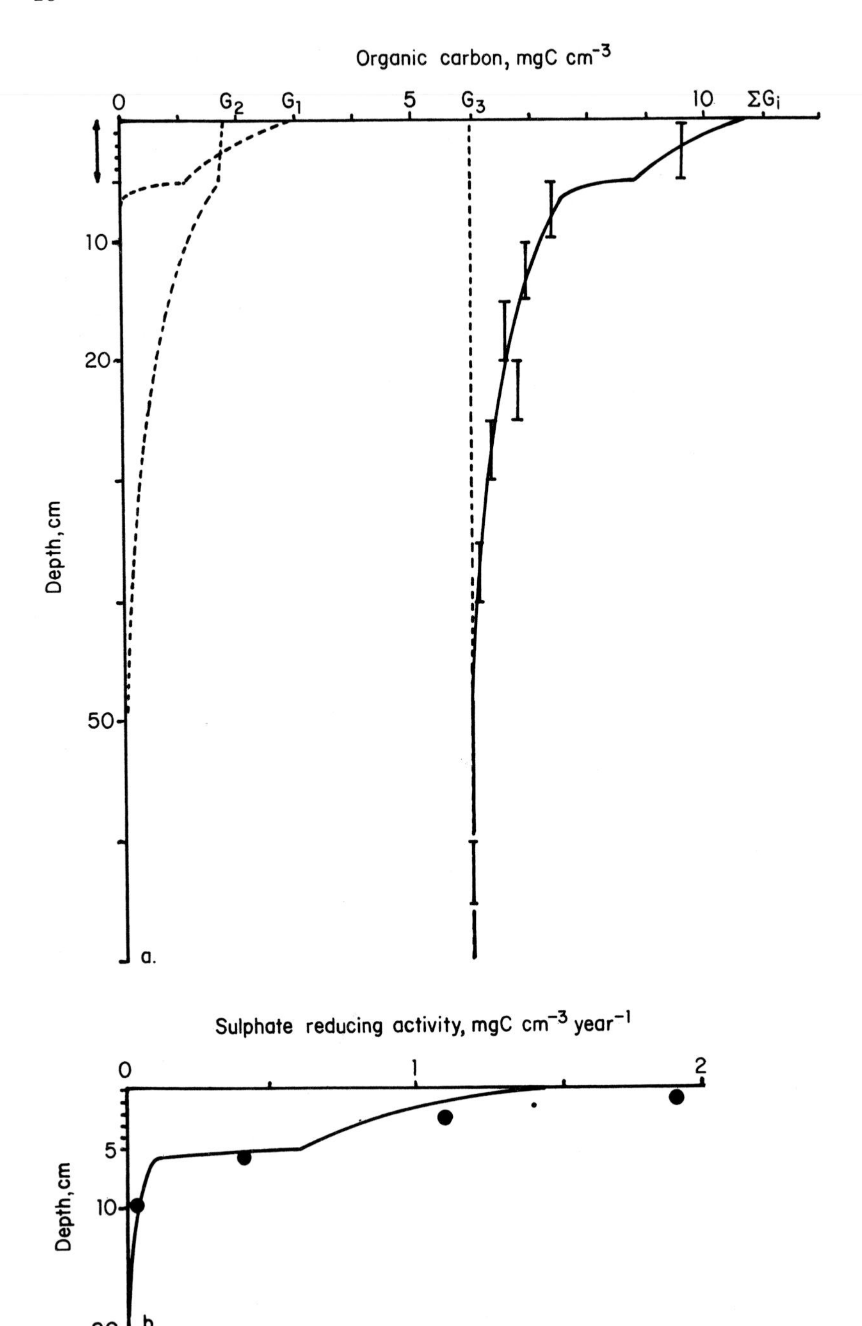

Fig. 2 (for legend see opposite).

Island Sound muddy sediments. The experimental organic carbon data of Benoit *et al.* [1979] have been fitted by the solution of a "3 G's" model constructed on the basis of data relative to the same site from Benoit *et al.* [1979], Turekian *et al.* [1980] and Aller *et al.* [1980]. In the zone of bioturbation, organic carbon displays a sharp decrease due to degradation of plankton-derived easily usable organic material (G_1, k_1). Below this zone, this material is rapidly and completely exhausted and the trend of the profile is dominated by degradation of a less easily decomposable class of organic matter (G_2, k_2). A last class of organic matter, possibly of terrestrial origin [Benoit *et al.*, 1979], seems to be entirely refractory (G_3, $k_3 \simeq 0$). The "characteristic depth" of decrease in organic matter content in the upper, bioturbated layer is given by the ratio $\sqrt{\frac{D_s}{k_1}}$ (relation 11) which can be estimated as 5.5 cm from data for D_s and k_1 from Aller *et al.* [1980] and Turekian *et al.* [1980] respectively. For the lower layer, the "characteristic depth" of organic matter decrease can be expressed as the ratio ω/k_2 (relation 10). By adjustment to the experimental profile, it can be estimated as 13 cm. With the value of 0.075 cm year^{-1} for ω [Benoit *et al.*, 1979] a k_2 value of 5.8×10^{-3} year^{-1}is implied.

It is seen in Fig. 2, that the pools G_2 and G_3 of less easily usable carbon dominate at all depths in the sediment profile. This contrasts with their only minor role for the benthic biological communities. The fluxes of organic carbon depositing on the sediment surface can be calculated according to relation (3) above:

$$J_{G_{io}} = [\sqrt{D_s k_i} + \omega] G_{io} \qquad (12)$$

Use of this relation for the three classes of organic material in the sediments of Long Island Sound gives:

Fig. 2 **a)** Calculated depth distribution of organic carbon in the sediments of Long Island Sound, according to a "3 G's" model using the data from Turekian *et al.* [1980] and Aller *et al.* [1980] (see text). The vertical bars indicate the organic carbon concentrations measured by Benoit *et al.* [1979]. The double arrow shows the depth of the bioturbation layer. **b)** Calculated values of organotrophic activities compared with measured sulphate reduction rates of Aller and Yingst [1980] in Long Island Sound sediments.

$$J_{G_{1o}} = 73 \text{ gC m}^{-2} \text{ year}^{-1}$$
$$J_{G_{2o}} = 6.5$$
$$J_{G_{3o}} = 4.5$$

$$J_{G_o} = 84 \text{ gC m}^{-2} \text{ year}^{-1}$$

showing that easily degraded plankton-derived organic matter constitutes 87% of the total input of organic carbon to the sediment.

For comparison, Table 3 gathers a selection of values of the flux of depositing organic matter on the sediments of various aquatic environments, as determined by direct (sediment trap) or indirect (balance calculations or measurements of benthic gross organotrophic activity) methods. The range is from about 1 gC m^{-2} $year^{-1}$ (0.25×10^{-9} mmole C cm^{-2} sec^{-1}) for deep-sea environments to near 200 gC m^{-2} $year^{-1}$ (50×10^{-9} mmoles C cm^{-2} sec^{-1}), in estuaries or eutrophic fresh water systems.

The "multi G's" model just discussed is also quite consistent with the experimental data on vertical distribution of heterotrophic bacterial biomass and activity in sediments [Vosjan and Olańczuk-Neyman, 1977; Goldhaber *et al.*, 1977; Aller and Yingst, 1980]. Most of these data show a rapid decrease in the upper few cm, with residual values decreasing more slowly in the lower layer, in good agreement with the relation obtained with the model by combining relations (6), (10) and (11). Figure 2b shows the expected rates of heterotrophic activity in Long Island Sound sediments, according to the model presented. They compare very well with the measurements of sulphate reduction rates reported by Aller and Yingst [1980] in the same site (sulphate reduction has been shown to be the most important heterotrophic process in these sediments).

Consumption of Oxidants and Redox Profiles in Sediments

Redox Potential and Oxidant Consumption

The microbial degradation of organic matter involves the consumption of an equivalent amount of mineral oxidants, either directly in the case of respiratory metabolisms, or indirectly in the case of fermentative metabolisms, the reduced products of which (organic acids, alcohols, H_2 etc.) have to be further oxidized by respiratory metabolisms. Oxidants used in microbial metabolism are oxygen, manganese oxides, nitrate and nitrite, ferric oxides, sulphate and carbon dioxide. Organic matter degradation within the sedimentary column causes a

TABLE 3

Flux of organic matter depositing on the sediments of some aquatic ecosystems, as estimated by various methods:
a) sediment trap b) total balance of organic matter cycling
c) analysis of bicarbonate concentration profiles in pore water

Location	Depth of the water column (m)	Flux of organic carbon (g C m^{-2} yr^{-1})	Method	Authors
Open Ocean (global estimate)	2000	1.2 - 2.6	b	Degens and Mopper [1976]
Harrington Sound (Bermuda)	25	12	c	Thorstenson and Mackenzie [1974]
Long Island Sound (USA)	14	6	b	Riley [1956]
		8.4	see text	This paper
Chesapeake Bay (USA)	15	32	b	Biggs and Flemer [1971]
	5	62	b	
Coastal North Sea (Belgium)	35	72	b	Nihoul and Polk [1977]
	15	160	b	
Loch Thurnaig (UK)	20 - 30	30	a	Davies [1974]
Lake Kinneret (Israel)	25	64	a	Serruya [1977]
Lake of Lucern (Switzerland)	43	67	a	Bloesch [1977]
Rotsee (Switzerland)	9	141	a	

depletion of these oxidants (X_i) and an accumulation of the corresponding reduced species (Y_i).

The concept and measurement of redox potential in natural environments have been discussed by several authors [Baas Becking *et al.*, 1960; Stumm, 1966; Thorstenson, 1970; Billen, 1978*b*]. It has been stated that the concept of redox potential in natural environments is meaningful owing to the fact that an internal thermodynamic equilibrium is reasonably approached within the subsystem formed by the main mineral redox species (A_i, Y_i) involved in the energy yielding metabolism of microorganisms. The redox potential (Eh) is defined with respect to this subsystem only and does not take into account the presence of highly reduced organic matter. It characterizes only the availability of oxidants susceptible to use by microbial respiration. Direct measurements of Eh in sediments with a platinum electrode must be interpreted with caution but can provide valuable relative indications [Whitfield, 1969; Bågander and Niemislö, 1978].

Model of Oxidants Distribution

As defined above, the redox potential of sediments can be viewed as the result of microbial metabolism. Organotrophic metabolisms generate a flux of electrons to the subsystem formed by mineral redox couples, while chemolithotrophic metabolism tends to restore the internal thermodynamic equilibrium by oxidizing reduced mineral species at the expense of oxidized ones, when this is thermodynamically possible. Knowing the rates and distribution of organotrophic activity within the sedimentary column, and the mixing processes to which all oxidants and their reduced forms are subject it is possible to calculate the vertical profiles of all oxidants and of the redox potential. This is the principle of a general idealized, redox model of marine sediments proposed by Billen and Verbeustel [1980]. Their model, however, was based on complete internal thermodynamic equilibrium, including nitrogen species, which is rather unrealistic. Moreover, only one mixing coefficient for both solid and dissolved species was considered. An improved version of this redox model, based on data collected in the sandy sediments of the North Sea, has been presented by Billen [in press]. It considers the following equilibria:

$$O_2 + 4e^- + 4H^+ \rightleftharpoons 2H_2O$$

$$\begin{cases} MnO_2 + 2e^- + 4H^+ \rightleftharpoons Mn^{2+} + 2H_2O \\ Mn^{2+} + HCO_3^- \rightleftharpoons MnCO_3 + H^+ \end{cases}$$

$$\begin{cases} Fe(OH)_3 + 1e^- + 3H^+ \rightleftharpoons Fe^{2+} + 3H_2O \\ Fe^{2+} + HCO_3^- \rightleftharpoons FeCO_3 + H^+ \end{cases}$$

$$SO_4^{2-} + 8e^- + 9H^+ \rightleftharpoons HS^- + 4H_2O$$
$$\{HS^- + FeCO_3 \rightleftharpoons FeS + HCO_3^-$$
$$HCO_3^- + 8e^- + 9H^+ \rightleftharpoons CH_4 + 3H_2O$$

Nitrate is not considered to be at equilibrium with respect to the other redox couples. It is assumed to be produced from ammonium through nitrification above a critical value of redox potential [Billen, 1975] and to be reduced into dinitrogen through denitrification below this potential. Organic material is assumed to be degraded according to first order (one G) kinetics and the resulting organotrophic activity causes an electron flux which is absorbed by the various oxidants according to the reactions listed above.

The redox profiles, theoretically predicted by this model for two different values of the input of fresh organic matter to marine sediments, are shown in Fig. 3a and 3b. Figure 3a corresponds to the situation termed "suboxic diagenesis" by Froelich *et al.* [1979], in which the organic matter input to the sediment is low enough with respect to the diffusion flux of oxidants so that only oxygen consumption, manganese-reduction, denitrification and ferrireduction are involved in organic matter degradation. Figure 3b on the other hand shows a situation of "anoxic diagenesis", where oxygen, manganese oxide, nitrate and ferric oxides are rapidly exhausted and sulphate reduction dominates organotrophic activity. Intermediate situations have been experimentally described by Sørensen *et al.* [1979].

The vertical profiles calculated by the equilibrium model are of course idealized. Moreover, they depend strongly on the values chosen for the various parameters (D_i, D_s, k). Nevertheless, they display the same general trends as numerous experimental observations.

(i) Oxygen concentration in organically rich sediments has been measured by means of microelectrodes by Revsbech *et al.* [1980*a* and *b*]. Their results indicate that in these sediments, oxygen does not penetrate deeper than a few millimetres below the water sediment interface. In organically poor sediments from Nova Scotia (Canada), however, Kepkay and Novitsky [1980] using a micro-Winkler procedure found measurable concentrations of oxygen down to 40 cm depth.

(ii) Manganese chemistry in marine sediments has been studied by numerous authors [Lynn and Bonatti, 1965; Robbins and Callender, 1974; Li *et al.*, 1969; Bender, 1971; Van der Weijden *et al.*, 1970; Michard, 1971; Calvert and Price, 1972; Troup and Bricker, 1974]. The most striking feature is the contrast between an upper layer of the sediments where the solid phase is enriched in manganese because of Mn^{++} oxidation and preciptation, and the lower layer with a lower manganese content in

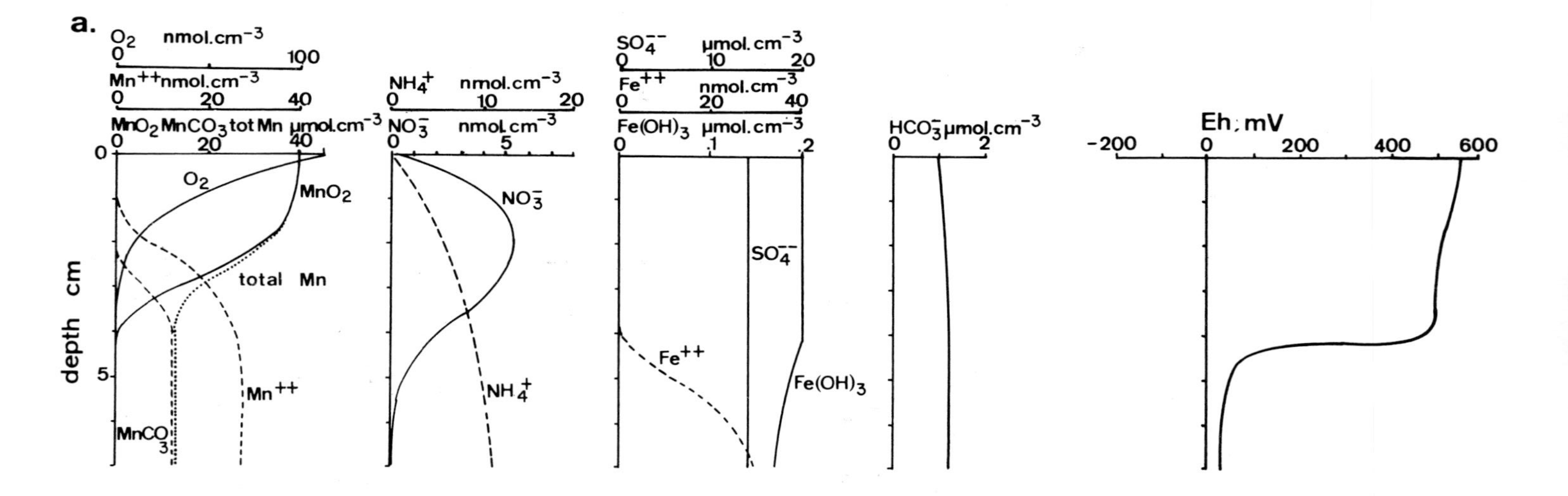

Fig. 3a For legend see page 34.

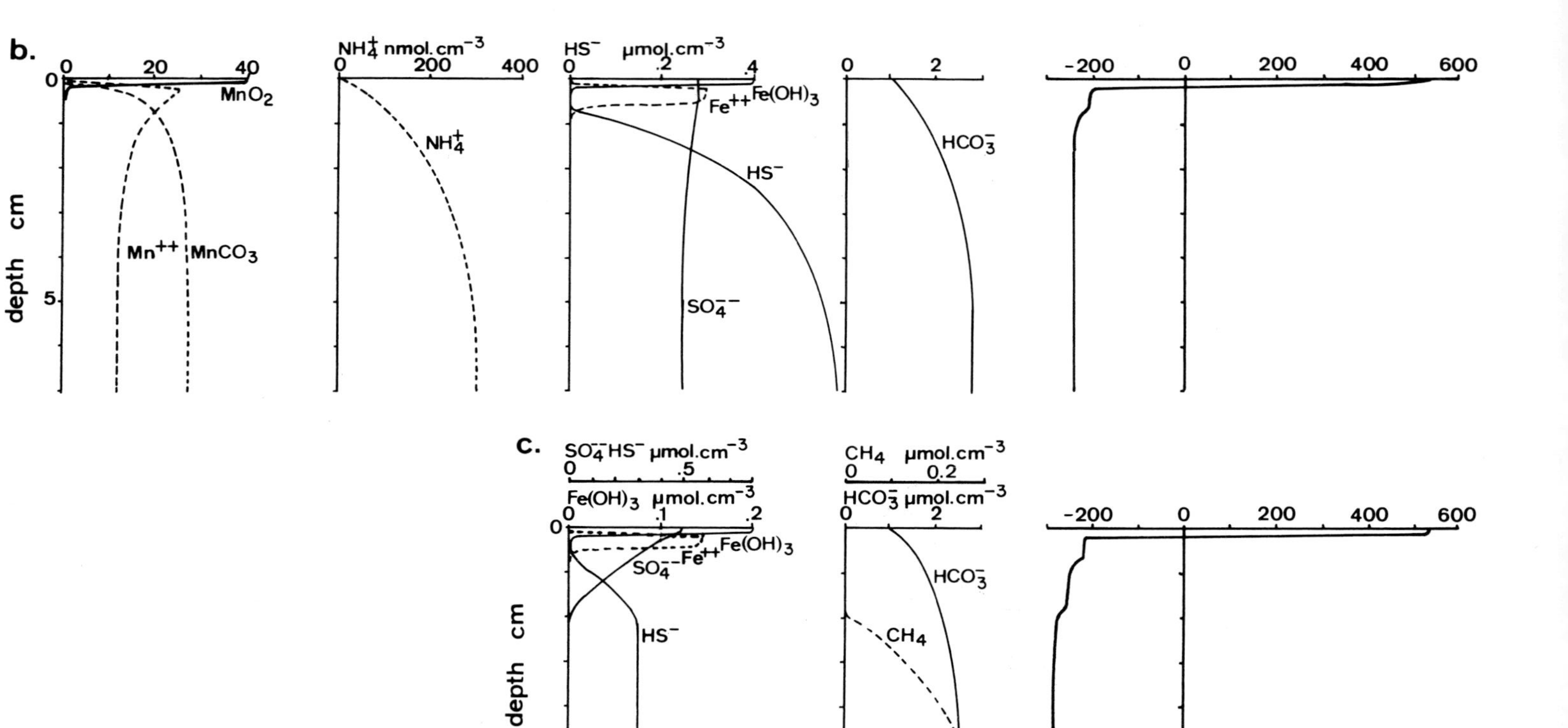

Fig. 3b and 3c For legend see page 34.

the solid phase but a higher concentration of dissolved manganese.
(iii) Nitrate concentration profiles with a maximum at a few centimetres depth result from nitrification in the upper layer and denitrification in the lower layer of most sediments as discussed by Vanderborght and Billen [1975] (see also below). Significant concentrations of nitrate are restricted to the upper oxidized layers.
(iv) Iron distribution in the sedimentary column shows many similarities to that of manganese [Bricker and Troup, 1975; Troup and Bricker, 1975]. Dissolved iron occurs at lower concentration in the reducing layers of organically rich sediments, because of the insolubility of iron sulphides. The redox transition between oxidized and reduced iron species in the solid phase is quite apparent in most sediments as a very sharp colour change from brown to grey or black occurring at a potential close to 100 mV at pH 7 [Pearsall and Mortimer, 1939; Mortimer, 1941, 1942; Fenchel, 1969]. When submitted to similar mixing conditions, organically poor sediments display this redox transition at deeper depth than sediments with higher organic input. The seasonal variations of organotrophic activity in sediments primarily resulting from temperature

---

Fig. 3 Calculated vertical profiles of mineral redox species in sediments according to thermodynamic equilibrium model [Billen, in press] for two different values of the flux of organic matter depositing on the sediment surface. ($D_i = 10^{-4}$ cm^2 sec^{-1}; $D_s = 10^{-7}$ cm^2 sec^{-1}; $k = 3 \times 10^{-8}$ sec^{-1}).

a) "sub-oxic diagenesis": case of a marine sediments with a flux of organic matter of 8×10^{-9} mmoles cm^{-2} sec^{-1} (30 gC m^{-2} year^{-1}) (overlying water containing 230 mM oxygen, 0 nitrate, 28 mM sulphate and 2 mM bicarbonate; upper sediment containing 40 µmole cm^{-3} manganese and 200 µmole cm^{-3} reactive iron; porosity = 0.5).

b) "anoxic diagenesis": case of a marine sediment with a flux of organic matter of 10^{-7} mmoles cm^{-2} sec^{-1} (378 gC m^{-2} year^{-1}) (same overlying water and sediment composition).

c) anoxic diagenesis with methane production: case of a fresh water sediment with a flux of organic matter of 10^{-7} mmoles cm^{-2} sec^{-1} (only 1 mM sulphate in the overlying water).

changes, result in variations of the depth of the brownish oxidized layer, as observed by several authors [Perkins, 1957; Billen and Verbeustel, 1980].
(v) In most organically rich marine sediments, sulphate reduction accounts for an important part of organotrophic activity. Consumption of sulphate in deeper, reducing layers creates a concentration gradient allowing diffusion of sulphates from the overlying water [Berner, 1964; Goldhaber *et al.*, 1977]. The sulphides produced rapidly precipitate ferrous iron in the form of greigite or mackinawite, which are further converted into pyrite [Berner, 1967, 1970; Goldhaber and Kaplan, 1974]. Dissolved sulphides accumulate in measurable quantities only after reactive iron has been precipitated as metal sulphides. Sulphide diffusing into the oxidized upper layer is reconverted into sulphate by chemolithotrophic sulphur-oxidizing bacteria [Kepkay and Novitsky, 1980].
(vi) Methane, mostly produced in sediments through acetate fermentation and carbon dioxide reduction, only builds up at appreciable concentrations in very reducing organically rich sediments, at depths where sulphate has been almost entirely exhausted [Martens and Berner, 1974, 1977; Reeburgh and Heggie, 1974; Barnes and Goldberg, 1976]. Recent data [Oremland and Taylor, 1978; Nedwell, this volume] show that although methane bacteria can be active in the sulphate reduction layer, competition with sulphate reducing bacteria for substrates strongly limits their activity. Furthermore, methane produced in the sulphate reduction layer or diffusing from lower layers can be oxidized by sulphate reducing bacteria [Martens and Berner, 1977]. These processes explain the differences in methane distribution between marine and lacustrine sediments [Reeburgh and Heggie, 1977]. The lower sulphate concentration of fresh water results in a more rapid exhaustion of sulphate from the interstitial water and in the occurrence of an active methane production layer at shallow depth, as illustrated by the model in Fig. 3c.

Redox Potential as a Function of Organic Input to the Sediments

Pearson and Stanley [1979] have recently used the measurement of redox potential in the sediments of a sea loch as a means of assessing the effect of organic pollution by a paper mill. They experimentally related the Eh reached in depth in the sedimentary column to the input of organic material to the sediments. Such a relation can be theoretically deduced from the redox model discussed above, as shown in Fig. 4 for a set of values of the mixing coefficients, D_i and D_s.

These curves show the "buffering capacity" of the various redox couples present in sediment toward "titration" by depositing organic matter. For oxidants

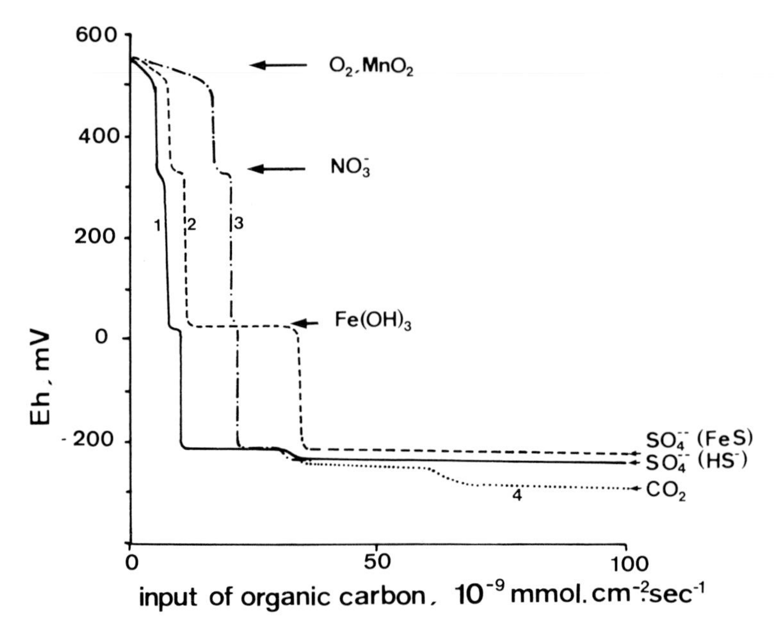

Fig. 4 Calculated relationship between the minimum redox potential reached in depth in sediments and the input of organic carbon depositing, according to a thermodynamic equilibrium model, for different values of the mixing coefficients

1) $D_i = 10^{-4}$ cm^2 sec^{-1} $D_s = 10^{-7}$ cm^2 sec^{-1} sea water
2) $D_i = 10^{-4}$ cm^2 sec^{-1} $D_s = 5 \times 10^{-6}$ cm^2 sec^{-1} sea water
3) $D_i = 10^{-4}$ cm^2 sec^{-1} $D_s = 10^{-8}$ cm^2 sec^{-1} sea water
4) $D_i = 10^{-4}$ cm^2 sec^{-1} $D_s = 10^{-7}$ cm^2 sec^{-1} fresh water

present in the solid phase this buffering capacity is closely dependent on the value of D_s, both because the mixing coefficient directly determines the "availability" of these oxidants within the sediment and because it determines the penetration of organic material down into the sediment, and thus the depth distribution of organotrophic activity, for a given flux of depositing organic matter.

Recycling of Nutrients

Mineralization of Organic Nitrogen and Phosphorus

Organic matter degradation generally results in the liberation of its constituent nitrogen and phosphorus in the form of ammonium and ortho-phosphate respectively. The C/N and C/P ratios of marine plankton are generally close to 6.6 and 106 respectively [Redfield, 1958], and degradation of organic carbon in the water column of either aerobic or anaerobic marine basins has been shown to produce nutrients according to the same ratios

[Redfield, 1958; Richards, 1965].

The situation may be slightly different in the case of organic matter degradation in sediments because of a longer previous history of organic matter decomposing there. Phosphorus and to a lesser extent nitrogen are known to be relatively more easily mineralized than carbon itself. Therefore a long residence in the water column or in upper layers of the sediments results in higher C/P and C/N ratios of the organic material susceptible to degradation. Thus Berner [1977] has shown that nitrogen and phosphorus are released according to C/N and C/P ratios of 8.8 and 141 respectively in the sulphate reducing layer of Long Island Sound sediments, while Aller and Yingst [1970] report a C/N ratio of 10 in the upper layers, and up to 24 in the lower layers of the sediments at a deeper site in the same environment.

Adsorption and Precipitation of Nutrients

Once released into the interstitial water of sediments, ammonium and phosphate ions can be adsorbed, most often reversibly, on various solid phases such as clay minerals, organic colloids, ferric oxides and hydroxides.

Berner [1976] and Schink and Guinasso [1978] have developed a general treatment for inclusion of adsorption processes in a diagenetic equation. When adsorption is rapid, reversible and follows a simple linear isotherm, the steady state diagenetic equation (2) for the concentration of the dissolved ion can be written:

$$0 = \frac{d}{dz}\left[D_s \frac{dC}{dz}\right] + \frac{1}{1+K}\frac{d}{dz}\left[D_i \frac{dC}{dz}\right] - \omega\frac{dC}{dz} + \left[\frac{1}{1+K}\right] R(z) \qquad (13)$$

where

D_s and D_i are the mixing coefficient of the solid particles and of the interstitial water respectively,
C is the concentration of dissolved species expressed per unit sediment volume,
R(z) is the net rate of production or consumption of the species expressed per unit sediment volume.
K is the adsorption constant supposed to be independent of depth and defined as the ratio between the concentration of adsorbed and dissolved species, both expressed per unit sediment volume.

Values of K for ammonium and phosphate in anoxic clay muds are generally within the range 1 to 2.5 [Rosenfeld, 1979; Krom and Berner, 1980]. Much lower values prevail of course in sands. Berner [1974, 1979, 1980*a*] has presented steady state models of ammonium distribution in the anoxic layer of marine sediments, taking into account adsorption and ammonification, which is simply considered as proportional to organic matter degradation, itself obeying first order, one G, kinetics:

$$R_{ammonification} = \frac{1}{\beta} k\,G$$

where β is the C/N ratio of the organic material, G, being degraded.

The case of phosphorus is a little more complex even in the anoxic layers because of the possibility of precipitation into authigenic minerals such as apatite or vivianite [Berner, 1974, 1980*a*]. In this case the following term for the net rate of phosphate production has been used:

$$R_{PO_4^{3-}\ production} = \frac{1}{\gamma} k\,G - km[C - Ceq]$$

where γ is the C/P ratio of the organic material G being degraded,
and km[C - Ceq] is a simple linear rate law for phosphate dissolution and precipitation.

Redox Processes Affecting Phosphorus and Nitrogen Recycling

The models of ammonium and phosphate concentration, briefly discussed in the preceding section, are restricted to the processes occurring in reducing sediments below the zone of bioturbation. In oxidizing conditions other processes have to be considered, which can greatly influence not only the depth distribution of nutrient concentrations but also their vertical exchange with the overlying water.

The presence of ferric oxides in the oxidized layer of the sediment has often been considered as a major factor controlling phosphorus recycling because of the strong adsorption properties of these phases for phosphates (adsorption constants, K, from 30 to over 3000 have been measured for oxic sediments with respect to phosphates, in contrast with the values of 1 to 2 cited above for anoxic sediments [Krom and Berner, 1980]). It has been stated that the oxic layer of sediments, when it exists, can act as a trap for phosphates produced by benthic mineralization. When the oxic layer is eliminated because of oxygen depletion in the water column (as often occurs in lakes), the adsorption capacity of the sediment is destroyed and phosphates are released to the overlying water [Mortimer, 1941, 1942; Fillos, 1977; Lijklema, 1977; Banoub, 1977]. Controversy exists however, concerning the quantitative role of these processes on the fluxes of phosphate to the overlying water [Lee *et al.*, 1977] and, to the author's knowledge, no satisfactory diagenetic model of phosphates in sediments with oxic and anoxic layers has been presented.

The role of the oxidized upper sediment layer on nitrogen recycling is better understood. In this layer ammonium is generally actively oxidized to nitrate by nitrifying bacteria. Direct measurements in the sediments of the North Sea have shown that nitrification rates are closely correlated with ammonification rate and amount to 80% of it [Billen, in press]. The depth of the layer where nitrification is possible is therefore a major factor in determining under which form (ammonium or nitrate) nitrogen is released to the water column. Moreover, nitrates formed in this oxidized layer can diffuse into the reduced layer and be reduced into dinitrogen which is far less accessible for primary production. The extent of this loss of nitrogen is also determined, in a complex way, by the depth of the oxidized layer.

From observations of the seasonal variations of the depth of the nitrification layer, and of nitrification and denitrification rates, in the bottom of a marine lagoon near Ostend, Belgium, (Fig. 5a), Billen and Vanderborght [1978] have calculated the fluxes of nitrate across the water-sediment interface, and the integrated rate of nitrification and denitrification, by means of steady state diagenesis modelling. Their results, represented in Fig. 5b, indicate two different patterns of nitrate exchanges in the two different sedimentary zones of the lagoon (one sandy, the other muddy) because of the different physico-chemical conditions prevailing there. In the sandy zone nitrate diffusion across the sediment-water interface is always directed upwards. The corresponding flux follows the seasonal variations of nitrification, with a maximum in the summer and a minimum in the winter. The increase of nitrification rate within the nitrification layer prevails over the decrease of the depth of this layer. Denitrification, on the other hand, remains approximately constant. In the muddy zone, integrated nitrification is maximum in the winter and minimum in the summer owing to the small depth of the oxidized layer at this period of the year. Denitrification is also maximum in the winter, decreases more slowly than nitrification in the spring, and reaches very small values in the summer when very little nitrate is present. The resulting nitrate flux across the sediment-water interface is sometimes directed upwards, sometimes downwards. It is to be noted that the sediment behaves as a sink of nitrate for the water column precisely during the period of the spring phytoplankton bloom, whose development can be significantly limited by this process.

A more general, although idealized, model of nitrogen recycling in sediments, based on data collected in the sandy sediments of the North Sea, has been presented by Billen [in press]. It relates the flux of organic material deposited on the sediments to the release of ammonium- and nitrate-nitrogen to the overlying water. The results of this model are shown in Fig. 6 for the same values of the mixing coefficients as those used in the redox model of Fig. 4. It is seen that at low input of organic material to the sediments (what can be considered as a "low" flux of organic material depends on the mixing conditions prevailing in the sediment) most

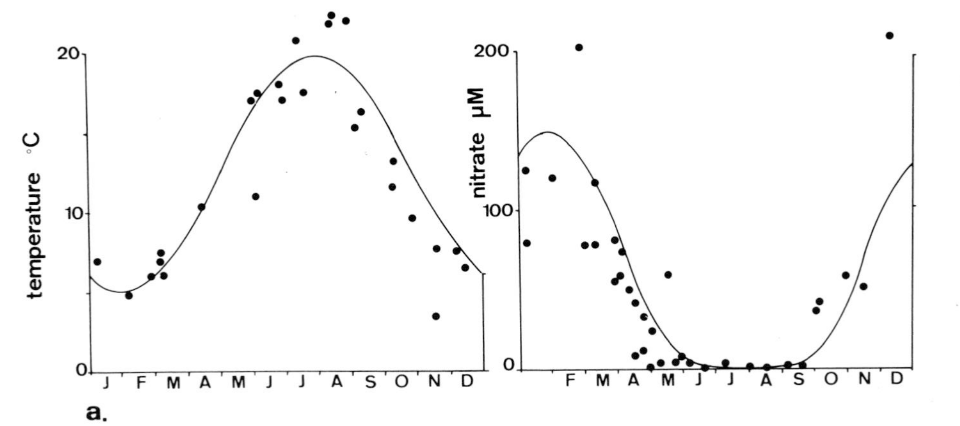

Fig. 5 Model of seasonal variations of nitrification and denitrification in the sediments of a marine lagoon near Ostend (Belgium). **a)** Observed seasonal evolution of temperature and nitrate concentration in the water of the lagoon (data from 1974 to 1976). Best fit by simple sinusoidal functions. **b)** Observed values of nitrification rate, depth of nitrification layer and first order constant of denitrification in the sandy (left) and muddy (right) sediments of the lagoon. Simulation by simple dependence on temperature. **c)** Seasonal evolution of the integrated rate of nitrification and denitrification and of the flux of nitrate across the water-sediment interface in the sandy (left) and muddy (right) sediments of the lagoon, as calculated according to the diagenetic model of Vanderborght and Billen [1975] using the values of the parameters shown in Fig. 5a and b ($D_i = 2 \times 10^{-4}$ cm^2 sec^{-1} for sandy and 10^{-4} cm^2 sec^{-1} for muddy sediments)

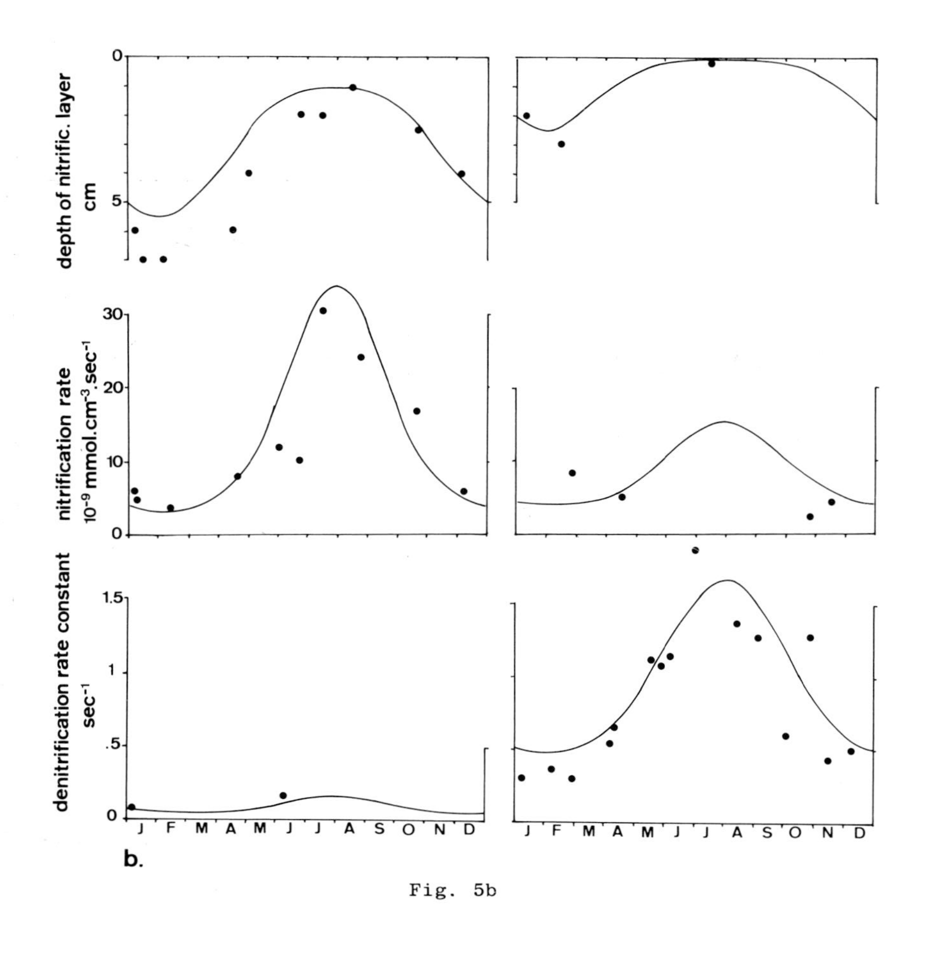
depth of nitrific. layer
cm
0
5
nitrification rate
10^{-9} mmol.cm^{-3}.sec^{-1}
30
20
10
0
denitrification rate constant
sec^{-1}
1.5
1
.5
0
J F M A M J J A S O N D
J F M A M J J A S O N D
b.

Fig. 5b

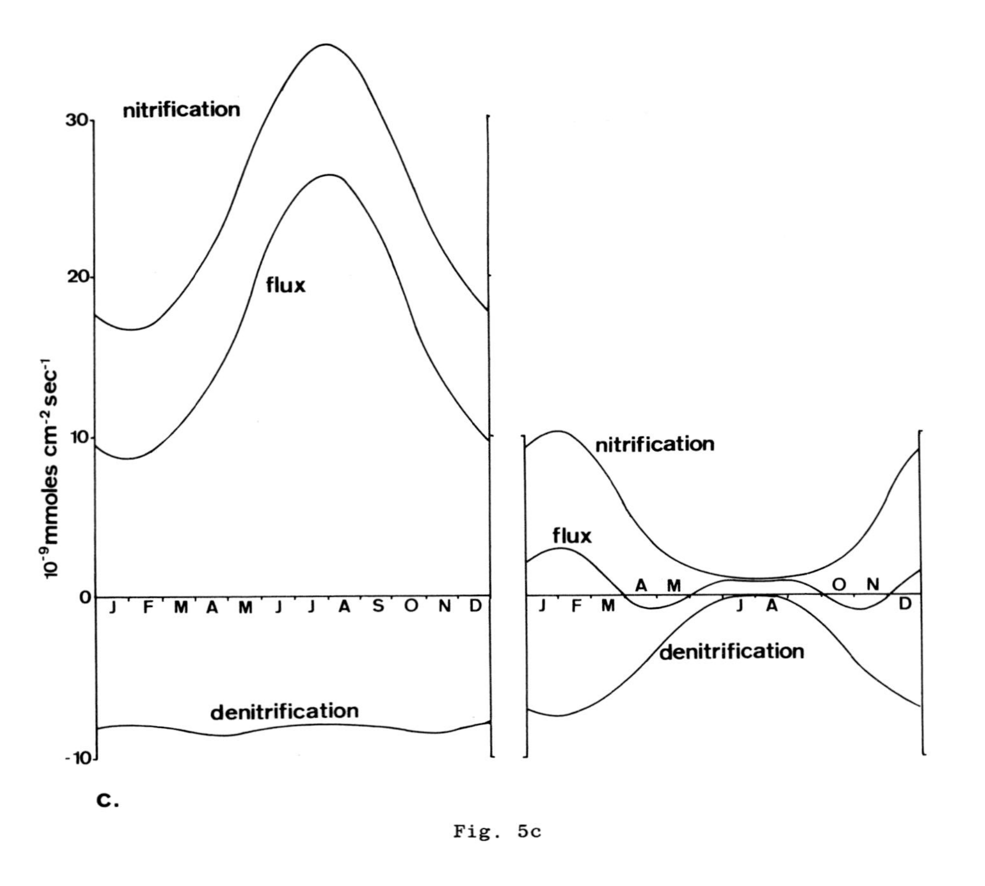
nitrification
flux
denitrification
30
20
10
0
-10
10^-9 mmoles cm^-2 sec^-1
J F M A M J J A S O N D
nitrification
flux
denitrification
J F M A M J A O N D
c.

Fig. 5c

of the nitrogen recycling occurs as nitrate. With increasing input of organic matter, nitrifying activity is restricted to a shallower and shallower upper layer (Fig. 6a), vertically integrated nitrification reaches a maximum, and ammonium release becomes more important. The last process prevails at high organic input (Fig. 6b). Denitrification, being dependent on nitrates formed in the nitrification layer, reaches a plateau above a certain input of organic material. Paradoxically it implies that the relative value of denitrification in the overall nitrogen cycle is maximum at an intermediate input of depositing material and decreases at higher inputs.

Conclusion

The purpose of this chapter was to show that even extremely idealized diagenetic models, taking into account the basic physical, chemical and microbiological processes affecting organic matter and nutrients in sediments, can account for the major trends of the observed behaviour of these substances in the sedimentary column. Modelling is not an end in itself, but it constitutes a powerful tool for testing postulated mechanisms, for making up quantitative balances of material cycling, or for investigating the effect of some environmental parameters on the dynamics of the sedimentary system.

Modelling, however, cannot work miracles. The value of the results it provides always reflects the value of the data or hypothesis used. It is therefore important to use them always in close combination with direct experimental methods for determining the mechanisms and the rate of microbial activity in sediments.

References

Allen, J.R.L. (1970). "Physical Processes of Sedimentation" Allen and Unwin Ltd, London.

Aller, R.C. (1978). Experimental studies of changes produced by deposit feeders on pore water, sediment and overlying water chemistry. *American Journal of Science* **278**, 1185-1234.

Aller, R.C., Benninger, L.K. and Cochran, J.K. (1980). Tracking particle-associated processes in nearshore environments by use of $^{234}Th/^{238}U$ disequilibrium. *Earth and Planetary Science Letters* **47**, 161-175.

Aller, R.C. and Cochran, J.K. (1976). $^{234}Th/^{238}U$ disequilibrium in nearshore sediment : particle reworking and diagenetic time scales. *Earth and Planetary Science Letters* **29**, 37-50.

Aller, R.C. and Yingst, J.Y. (1980). Relationships between microbial distributions and the anaerobic decomposition of organic matter in surface sediments of Long Island Sound (USA). *Marine Biology* **56**, 29-42.

Anikouchine, W.A. (1967). Dissolved chemical substances in compacting marine sediments. *Journal of Geophysical Research* **72**, 505-509.

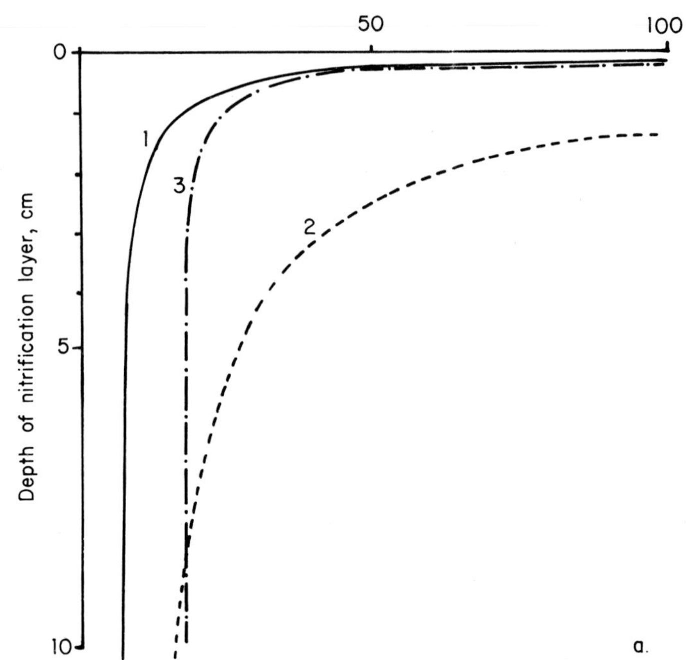

Fig. 6 Model of nitrogen recycling in marine sediments as a function of the input of organic material, for different values of the mixing coefficients.
1) $D_i = 10^{-4}$ cm^2 sec^{-1} $D_s = 10^{-7}$ cm^2 sec^{-1}
2) $D_i = 10^{-4}$ cm^2 sec^{-1} $D_s = 5 \times 10^{-6}$ cm^2 sec^{-1}
3) $D_i = 10^{-4}$ cm^2 sec^{-1} $D_s = 10^{-8}$ cm^2 sec^{-1}
C/N ratio = β = 6
a) Depth of the nitrification layer.

Arrhenius, G.O.S. (1963). Pelagic sediments. In "The Sea" (Ed. M.N. Hill) Vol.III, pp.658-727. Wiley, New York.
Baas Becking, L.G.M., Kaplan, I.R. and Moore, D. (1960). Limits of the natural environment in terms of pH and oxidation-reduction potential. *Journal of Geology* **68**, 243-284.
Bågander, L.E. and Niemistö, L. (1978). An evaluation of the use of redox measurements for characterizing recent sediments. *Estuarine and Coastal Marine Science* **6**, 127-134.
Bagnold, R.A. (1966). An approach to the sediment transport problem from general physics. *Geological Survey Professional Paper* 422-1-USGPO.
Banoub, M.W. (1977). Experimental investigation on the release of phosphorus in relation to iron in freshwater, mud system. In "Interactions Between Sediment and Freshwater" Proceedings of a symposium held in Amsterdam. (Ed. H.L. Golterman), pp.324-330. Dr W. Junk, B.V. Publishers, The Hague.
Barnes, R.O. and Goldberg, E.D. (1976). Methane production and consumption in anoxic marine sediments. *Geology* **4**, 297-300.

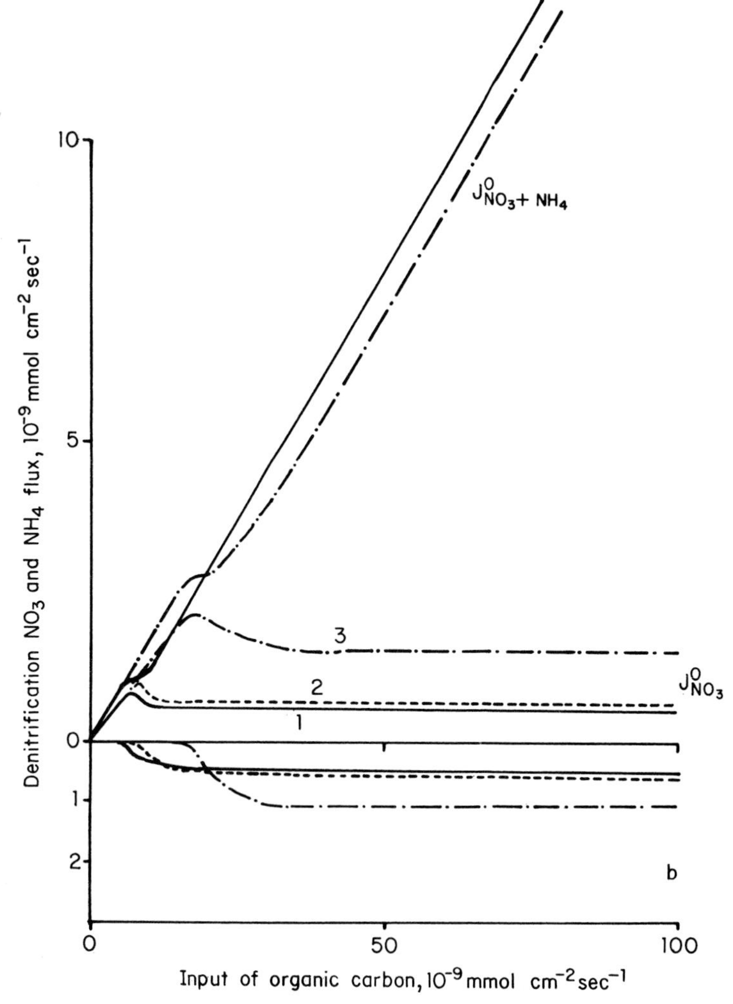

Fig. 6 (cont.) b) Fluxes of nitrate and total mineral nitrogen across the water sediment interface and integrated rate of denitrification.

Bender, M.L. (1971). Does upward diffusion supply the excess manganese in pelagic sediments? *Journal of Geophysical Research* **76**, 4212-4215.

Benoît, G.S., Turekian, K.K. and Benninger, L.K. (1979). Radiocarbon dating of a core from Long Island Sound. *Estuarine and Coastal Marine Science* **9**, 171-180.

Berger, W.H. and Heath, G.R. (1968). Vertical mixing in pelagic sediments. *Journal of Marine Research* **26**, 134-143.

Berner, R.A. (1964). An idealized model of dissolved sulphate distribution in recent sediments. *Geochimica et Cosmochimica Acta* **28**, 1497-1503.

Berner, R.A. (1967). Thermodynamic stability of sedimentary iron sulphides. *American Journal of Science* **265**, 773-785.
Berner, R.A. (1970). Sedimentary pyrite formation. *American Journal of Science* **268**, 1-23.
Berner, R.A. (1971). "Principles of Chemical Sedimentology" McGraw Hill, New York.
Berner, R.A. (1974). Kinetic models for the early diagenesis of nitrogen, sulphur, phosphorus and silicon in anoxic marine sediments. In "The Sea" (Ed. E.D. Goldberg) Vol.V, pp.427-450. Wiley, New York.
Berner, R.A. (1975). Diagenetic models of dissolved species in the interstitial waters of compacting sediments. *American Journal of Science* **275**, 88-96.
Berner, R.A. (1976). Inclusion of adsorption in the modelling of early diagenesis. *Earth and Planetary Science Letters* **29**, 333-340.
Berner, R.A. (1977). Stoichiometric models for nutrient regeneration in anoxic sediments. *Limnology and Oceanography* **22**, 781-786.
Berner, R.A. (1978). Sulphate reduction and the rate of deposition of marine sediments. *Earth and Planetary Science Letters* **37**, 492-498.
Berner, R.A. (1979). Kinetics of nutrient regeneration in anoxic marine sediments. In "Origin and Distribution of the Elements" (Ed. L.H. Ahrens) Vol.2, pp.279-292. Pergamon Press, Oxford.
Berner, R.A. (1980*a*). "Early Diagenesis - A Theoretical Approach" Princeton University Press, Princeton, N.J.
Berner, R.A. (1980*b*). A rate model for organic matter decomposition during bacterial sulphate reduction in marine sediments. In "Biogéochemie de la Matière Organique à l'Interface Eau-Sédiment marin" pp.35-45. Colloque International du Centre National de la Recherche Scientifique, n° 293. CNRS, Paris.
Biggs, R.B. and Flemer, D.A. (1971). The flux of particulate carbon in an estuary. *Marine Biology* **12**, 11-17.
Billen, G. (1975). Nitrification in the Scheldt estuary (Belgium and The Netherlands). *Estuarine and Coastal Marine Science* **3**, 79-89.
Billen, G. (1976). A method for evaluating nitrifying activity in sediments by dark ^{14}C-bicarbonate incorporation. *Water Research* **10**, 51-57.
Billen, G. (1978*a*). A budget of nitrogen recycling in North Sea sediments off the Belgian coast. *Estuarine and Coastal Marine Science* **7**, 127-246.
Billen, G. (1978*b*). The dependence of the various kinds of microbial metabolism on the redox state of the medium. In "Biogeochemistry of Estuarine Sediments" pp.254-261. Proceedings of a UNESCO/SCOR workshop. UNESCO, Paris.
Billen, G. An idealized model of nitrogen recycling in marine sediments. *American Journal of Science*. (in press).
Billen, G., Joiris, C., Wijnant, J. and Gillain, G. (1980). Concentration and microbial utilization of small organic molecules in the Scheldt estuary, the Belgian coastal zone of the North Sea and the English Channel. *Estuarine and Coastal Marine Science* **11**, 279-294.
Billen, G. and Vanderborght, J.P. (1978). Evaluation of the exchange fluxes of materials between sediments and overlying waters from direct measurements of bacterial activity and mathematical analy-

sis of vertical concentration profiles in interstitial waters. In "Biogeochemistry of Estuarine Sediments" pp.115-165. Proceedings of UNESCO/SCOR workshop. UNESCO, Paris.

Billen, G. and Verbeustel, S. (1980). Distribution of microbial metabolisms in natural environments displaying gradients of oxidation-reduction conditions. In "Biogéochemie de la Matière Organique à l'Interface Eau-Sediment marin" pp.291-301. Colloque International du Centre National de la Recherche Scientifique, n° 293. CNRS, Paris.

Bloesch, J. (1977). Sedimentation rates and sediment cores in two Swiss lakes of different trophic state. In "Interactions Between Sediments and Freshwater" (Ed. H.L. Colterman), pp.65-71. Proceedings of a symposium held in Amsterdam. Dr W. Junk, B.V. Publishers, The Hague.

Bloesch, J. (1978). A critical review of the sedimentation trap technique. UNESCO/Workshop on the assessment of particulate matter contamination in rivers and lakes. 13-17 November, 1978. Budapest.

Bricker, O.W. and Troup, B.N. (1975). Sediment-water exchange in Chespeake Bay. In "Estuarine Research: Chemistry, Biology and the Estuarine System" Vol.1, pp.3-27, Academic Press, New York.

Calvert, J.E. and Price, N.B. (1972). Diffusion and reaction of dissolved manganese in the pore waters of marine sediments. *Earth and Planetary Science Letters* **16**, 245-249.

Carey, C. (1967). Energetics of the benthos of Long Island Sound. I. Oxygen utilization of sediments. *Bulletin of the Bingham Oceanographic Collection* **19**, 136-144.

Christensen, D. and Blackburn, T.H. (1980). Turnover of tracer (^{14}C, ^{3}H labelled) alanine. *Marine Biology* **58**, 97-103.

Dales, R.P. (1961). Oxygen uptake and irrigation of the burrow by three ferebillid polychaetes: Eupolymnia, Thelepus and Neamphitrite. *Physiological zoology* **34**, 306-311.

Dapples, E.C. (1942). The effect of macro-organisms upon nearshore marine sediments. *Journal of Sedimentary Petrology* **12**, 118-126.

Davies, J.M. (1974). Energy flow through the benthos in a Scottish sea loch. *Marine Biology* **31**, 353-362.

Degens, E.T. and Mopper, K. (1975). Early diagenesis of organic matter in marine soils. *Soil Science* **119**, 65-72.

Degens, E.T. and Mopper, K. (1976). Factors controlling the distribution and early diagenesis of organic material in marine sediments. In "Chemical Oceanography" (Eds. J.P. Riley and R. Chester) Vol.6, pp.60-113. Academic Press, New York.

Duursma, E.K. and Gross, M.G. (1971). Marine sediments and radioactivity. In "Radioactivity in the Marine Environment" pp.147-160. National Academy of Science, Washington.

Fanning, K.A. and Pilson, M.E.Q. (1974). The diffusion of dissolved silica out of deep-sea sediments. *Journal of Geophysical Research* **79**, 1293-1297.

Fenchel, T. (1969). The ecology of marine microbenthos. *Ophelia* **6**, 1-182.

Fillos, J. (1977). Effects of sediments on the quality of the overlying water. In "Interactions Between Sediments and Fresh Water"

(Ed. H.L. Golterman), pp.266-271. Proceedings of a symposium held in Amsterdam. Dr W. Junk, B.V. Publishers, The Hague.

Froelich, P.N., Klinkhammer, G.P., Bender, M.L., Luedtke, N.A., Heath, G.R., Cullen, D., Dauphin, P., Hammond, D., Hartman, B. and Maynard, V. (1979). Early oxidation of organic matter in pelagic sediments of the eastern equatorial Atlantic: suboxic diagenesis. *Geochimica et Cosmochimica Acta* **43**, 1075-1090.

Goldberg, E.D. and Koide, M. (1962). Geochronological studies of deep-sea sediments by the thorium-ionium method. *Geochimica et Cosmochimica Acta* **26**, 417-450.

Goldhaber, M.B., Aller, R.C., Cochran, J.K., Rosenfeld, J.F., Martens, C.S. and Berner, R.A. (1977). Sulphate reduction, diffusion and bioturbation in Long Island Sound sediments: report of the FOAM Group. *American Journal of Science* **227**, 193-237.

Goldhaber, M.B. and Kaplan, I.R. (1974). The sulphur cycle. In "The Sea" (Ed. E.D. Goldberg) Vol.V, pp.569-655. Wiley, New York.

Gordon, D.C. (1966). The effects of the deposit feeding polychaete *Pectinaria gouldii* on the intertidal sediments of Barnstable Harbor. *Limnology and Oceanography* **11**, 327-332.

Guinasso, N.L. and Schink, D.R. (1975). Quantitative estimate of biological mixing rates in abyssal sediments. *Journal of Geophysical Research* **80**, 3032-3043.

Hargrave, B.T.H. (1973). Coupling carbon flow through some pelagic and benthic communities. *Journal of Fisheries Research Board of Canada* **30**, 1317-1326.

Hargrave, B.T. and Burns, N.M. (1979). Assessment of sediment trap collection efficiency. *Limnology and Oceanography* **24**, 1124-1136.

Imboden, D.M. (1975). Interstitial transport of solutes in non-steady state accumulating and compacting sediments. *Earth and Planetary Science Letters* **27**, 221-228.

Iturriaga, R. (1979). Bacterial activity related to sedimenting particulate matter. *Marine Biology* **55**, 157-169.

James, A. (1974). The measurement of benthal respiration. *Water Research* **8**, 955-959.

Jørgensen, B.B. (1978*a*). A comparison of methods for the quantification of bacterial sulphate reduction in coastal marine sediments. I. Measurement with radiotracer techniques. *Geomicrobiology Journal* **1**, 11-27.

Jørgensen, B.B. (1978*b*). A comparison of methods for the quantification of bacterial sulphate reduction in coastal marine sediments. II. Calculation from mathematical models. *Geomicrobiology Journal* **1**, 29-47.

Kepkay, P.E. and Novitsky, J.A. (1980). Microbial control of organic carbon in marine sediments: coupled chemoautototrophy and heterotrophy. *Marine Biology* **55**, 261-266.

Krishnaswami, S., Benninger, L.K., Aller, R.C. and Von Damm, K.L. (1980). Atmospherically-derived radionuclides as tracers of sediment mixing and accumulation in near-shore marine and lake sediments: evidence from ^{7}Be, ^{210}Pb and ^{239,240}Pu. *Earth and Planetary Science Letters* **47**, 307-318.

Krom, M.D. and Berner, R.A. (1980). Adsorption of phosphate in anoxic marine sediments. *Limnology and Oceanography* **25**, 797-806.

Laughton, A.S. (1963). Microtopography. In "The Sea" (Ed. M.N. Hill) Vol.3, pp.437-472. Wiley, New York.

Lee, G.F., Sonzogni, W.L. and Spear, R.D. (1977). Significance of oxic versus anoxic conditions for lake Mendota sediment phosphorus release. In "Interactions Between Sediments and Fresh Water" (Ed. H.L. Golterman) pp.294-306. Proceedings of a symposium held in Amsterdam. Dr W. Junk, B.V. Publishers, The Hague.

Lerman, A. (1979). "Geochemical Processes: Water and Sediment Environments" Wiley, New York.

Li, Y.H., Bischoff, J. and Mathieu, G. (1969). The migration of manganese in the arctic basin sediment. *Earth and Planetary Science Letters* **7**, 265-270.

Li, Y.H. and Gregory, S. (1974). Diffusion of ions in sea water and in deep-sea sediments. *Geochimica et Cosmochimica Acta* **38**, 703-714.

Lijklema, L. (1977). The role of iron in the exchange of phosphate between water and sediments. In "Interactions Between Sediments and Fresh Water" (Ed. H.L. Golterman) pp.313-317. Proceedings of a symposium held in Amsterdam. Dr W. Junk, B.V. Publishers, The Hague.

Linch, D.L. and Cotnoir, L.J. (1956). The influence of clay minerals on the breakdown of certain organic substrates. *Proceedings of the Soil Science Society of America* **20**, 367-380.

Luedtke, N.A. and Bender, M.L. (1979). Tracer study of sediment-water interactions in estuaries. *Estuarine and Coastal Marine Science* **9**, 643-651.

Lynn, D.C. and Bonatti, E. (1965). Mobility of manganese in diagenesis of deep sea sediments. *Marine Biology* **3**, 457-474.

MacCaffrey, R.J., Myers, A.C., Davey, E., Morrisson, G., Bender, M., Luedtke, N., Cullen, D., Froelich, P. and Klinkhammer, G. (1980). The relation between pore water chemistry and benthic fluxes of nutrients and manganese in Narragansett Bay, Rhode Island. *Limnology and Oceanography* **25**, 31-44.

McDuff, R.E. and Ellis, R.A. (1979). Determining diffusion coefficient in marine sediments: a laboratory study of the validity of resistivity techniques. *American Journal of Science* **279**, 666-675.

Mangum, C.P. (1964). Activity patterns in metabolism and ecology of polychaetes. *Comparative Biochemistry and Physiology* **11**, 239-256.

Martens, C.S. and Berner, R.A. (1974). Methane production in the interstitial waters of sulphate-depleted marine sediments. *Science* **185**, 1167-1169.

Martens, C.S. and Berner, R.A. (1977). Interstitial water chemistry of anoxic Long Island Sound sediments. 1. Dissolved gases. *Limnology and Oceanography* **22**, 10-25.

Michard, G. (1971). Theoretical model for manganese distribution in calcareous sediment cores. *Journal of Geophysical Research* **76**, 2179-2186.

Mortimer, C.H. (1941). The exchange of dissolved substances between mud and water in lakes. *Journal of Ecology* **29**, 280-329.

Mortimer, C.H. (1942). The exchange of dissolved substances between mud and water in lakes. *Journal of Ecology* **30**, 147-201.

Nichols, F.H. (1974). Sediment turnover by a deposit-feeding polychaete. *Limnology and Oceanography* **19**, 945-950.

Nihoul, J.C.J. and Polk, P. (1977). Dynamiek van het ecosystem Noordzee. Modèle Mathématique la Mer du Nord (Projet Mer) Vol.7.

Service du Premier Ministre, programmation de la politique scientifique, Brussels.
Nozaki, Y., Cochran, J.K. and Turekian, K.K. (1977). Radiocarbon and ^{210}Pb distribution in submersible-taken deep-sea cores from project famous. *Earth and Planetary Science Letters* **34**, 167-173.
Oremland, R.S. and Taylor, B.F. (1978). Sulphate reduction and methanogenesis in marine sediments. *Geochimica et Cosmochimica Acta* **42**, 209-214.
Otsuki, A. and Hanya, T. (1972). Production of dissolved organic matter from dead green algae cells. I. Aerobic microbial decomposition. *Limnology and Oceanography* **17**, 248-264.
Pearsall, W.H. and Mortimer, C.H. (1939). Oxidation-reduction potentials in water-logged soils, natural waters and muds. *Journal of Ecology* **27**, 463-501.
Pearson, T.H. and Stanley, S.O. (1979). Comparative measurement of the redox potential of marine sediments as a rapid means of assessing the effect of organic pollution. *Marine Biology* **53**, 371-379.
Perkins, E.J. (1957). The blackened sulphide containing layer of marine soils, with special references to that found at Whitstable, Kent. *Annual Magazine of Natural History* **12**, 25-35.
Putnam, J.A. (1949). Loss of wave energy due to percolation in a permeable sea bottom. *Transaction of the American Geophysical Union* **30**, 349-356.
Redfield, A.C. (1958). The biological control of chemical factors in the environment. *American Scientist* **46**, 205-222.
Reeburgh, W.S. and Heggie, D.T. (1974). Depth distribution of gases in shallow water sediments. In "Natural Gases in Marine Sediments" (Ed. I.R. Kaplan) pp.27-45. Plenum, New York.
Reid, R.O. and Kajiura, K. (1957). On the damping of gravity waves over a permeable bed. *Transaction of American Geophysical Union* **38**, 662-666.
Revsbech, N.P., Jørgensen, B.B. and Blackburn, T.H. (1980*a*). Oxygen in the Sea bottom measured with a microelectrode. *Science* **207**, 1355-1356.
Revsbech, N.P., Sørensen, J., Blackburn, T.H. and Lomholt, J.P. (1980*b*). Distribution of oxygen in marine sediments measured with microelectrodes. *Limnology and Oceanography* **25**, 403-411.
Rhoads, D.C. (1963). Rates of sediment reworking by *Yoldia limatula* in Buzzards Bay (Massachussetts) and Long Island Sound. *Journal of Sedimentary Petrology* **33**, 723-727.
Rhoads, D.C. (1967). Biogenic reworking of intertidal and subtidal sediments in Barnstable Harbor and Buzzards Bay, Massachussetts. *Journal of Geology* **75**, 461-476.
Rhoads, D.C. (1973). The influence of deposit-feeding benthos on water turbidity and nutrient recycling. *Americal Journal of Science* **273**, 1-22.
Richards, F.A. (1965). Anoxic basins and fjords. In "Chemical Oceanography" (Eds. S.P. Riley and G. Skirrow) Vol.I, pp.611-645. Academic Press, New York.
Riedl, R.J., Huang, N. and Machan, R. (1972). The subtidal pump: a mechanism of interstitial water exchange by wave action. *Marine Biology* **13**, 210-221.
Riley, G.A. (1956). Oceanography of Long Island Sound, 1952-1954:

Production and utilization of organic matter. *Bulletin of the Bingham Oceanographic Collection* **15**, 324-344.
Rittenberg, S.C., Emery, K.O. and Orr, W.L. (1955). Regeneration of nutrients in sediments of marine basins. *Deep-Sea Research* **3**, 23-45.
Robbins, J.A. and Callender, E. (1974). Diagenesis of manganese in Lake Michigan sediments. Great Lakes Research Division, Contribution n° 177.
Robbins, J.A., Krezoski, J.R. and Mozley, S.C. (1977). Radioactivity in sediments of the great lakes: post-depositional redistribution by deposit-feeding organisms. *Earth and Planetary Science Letters* **36**, 325-333.
Rosenfeld, J.K. (1979). Ammonium adsorption in nearshore anoxic sediments. *Limnology and Oceanography* **24**, 356-364.
Rowe, G.T., Clifford, C.H., Smith, K.L.Jr and Hamilton, P.L. (1975). Benthic nutrient regeneration and its coupling to primary productivity in coastal waters. *Nature* **255**, 215-217.
Saunders, G.W. (1972). The transformation of artificial detritus in lake water. *Memorie del Instituto Italiano d'Idorbiologia* **29**, (suppl) 261-288.
Schink, D.R. and Guinasso, N.L. (1978). Redistribution of dissolved and adsorbed materials in abyssal marine sediments undergoing Biological Stirring. *American Journal of Science* **278**, 687-702.
Seitzinger, S., Nixon, S., Pilson, M.E.Q. and Burke, S. (1980). Denitrification and N_2O production in near-shore marine sediments. *Geochimica et Cosmochimica Acta* **44**, 1853-1860.
Serruya, C. (1977). Rates of sedimentation and resuspension in Lake Kinneret. In "Interactions Between Sediments and Fresh Water" (Ed. H.L. Golterman) pp.48-56. Proceedings of a symposium held in Amsterdam. Dr W. Junk, V.B. Publishers, The Hague.
Smith, K.L., Burns, K.A. and Teal, J.M. (1972). *In situ* respiration of benthic communities in Castle Harbor, Bermuda. *Marine Biology* **12**, 196-199.
Sørensen, J. (1978). Denitrification rates in a marine sediment as measured by the acetylene inhibition technique. *Applied and Environmental Microbiology* **36**, 139-143.
Sørensen, J., Jørgensen, B.B. and Revsbech, N.P. (1979). A comparison of oxygen, nitrate and sulphate respiration in coastal marine sediments. *Microbial Ecology* **5**, 105-115.
Steele, J.H., Munro, A.L.S. and Giese, G.S. (1970). Environmental factors controlling the episammic flora on beach and sublittoral sands. *Journal of the Marine Biological Association of the United Kingdom* **50**, 907-918.
Streeter, H.W. and Phelps, E.B. (1925). Study of the pollution and natural purification of the Ohio river. III. Factors concerned in the phenomena of oxidation and reaeration. Bulletin of the United States Public Health Service n° 146.
Stumm, W. (1966). Redox potential as an environmental parameter: conceptual significance and operational limitation. Proceeding of the International Water Pollution Research Conference (3rd Munich) **1**, 283-308.
Thorstenson, D.C. (1970). Equilibrium distribution of small organic molecules in natural waters. *Geochimica et Cosmochimica Acta* **34**, 745-770.

Thorstenson, D.C. and Mackenzie, F.T. (1974). Time variability of pore water chemistry in recent carbonate sediments. Devil's Hole, Harrington Sound, Bermuda. *Geochimica et Cosmochimica Acta* **38**, 1-18.

Toth, D.J. and Lerman, A. (1977). Organic matter reactivity and sedimentation rates in the ocean. *American Journal of Science* **277**, 465-485.

Troup, B.N. and Bricker, O.P. (1974). Processes affecting the transport of materials from continents to oceans. In "Marine Chemistry in the Coastal Environment" (Ed. T.M. Church) pp.133-151. Americal Chemical Society, Symposium series n° 18.

Turekian, K.K., Benoit, G.S. and Benninger, L.K. (1980). The mean residence time of plankton-derived carbon in a Long Island Sound sediment core: a correction. *Estuarine and Coastal Marine Sciences* **11**, 583.

Turekian, K.K., Cochran, J.K. and Demaster, D.J. (1978). Bioturbation in deep-sea deposits: rates and consequences. *Oceanus* **21**, 34-41.

Tzur, Y. (1971). Interstitial diffusion and advection of solute in accumulating sediments. *Journal of Geophysical Research* **76**, 4208-4211.

Vanderborght, J.P. and Billen, G. (1975). Vertical distribution of nitrate in interstitial water of marine sediments with nitrification and denitrification. *Limnology and Oceanography* **20**, 953-961.

Vanderborght, J.P., Wollast, R. and Billen, G. (1977). Kinetic models of diagenesis in disturbed sediments. I. Mass transfer properties and silica diagenesis. *Limnology and Oceanography* **22**, 787-793.

Van der Weijden, C.H., Schuiling, R.D. and Das, H.A. (1970). Some geochemical characteristics of sediments from the North Atlantic Ocean. *Marine Geology* **9**, 81.

Vosjan, J.H. and Olańczuk-Neyman, K. (1977). Vertical distribution of mineralization processes in a tidal sediment. *Netherlands Journal of Sea Research* **11**, 14-23.

Webb, J.E. and Theodor, J.L. (1968). Wave induced circulation in submerged sands through wave action. *Nature* **220**, 682-683.

Webb, J.E. and Theodor, J.L. (1972). Wave induced circulation in submerged sands. *Journal of the Marine Biological Association of the United Kingdom* **52**, 903-914.

Welte, D. (1973). Recent advances in organic geochemistry of humic substances and kerogen. In "Advances in Organic Geochemistry" (Eds. B. Tissot and F. Bieuner), pp.4-13. Technip, Paris.

Whitfield, M. (1969). Eh as an operational parameter in estuarine studies. *Limnology and Oceanography* **14**, 547-558.

Wieser, W. and Zech, M. (1976). Dehydrogenases as tools in the study of marine sediments. *Marine Biology* **36**, 113-122.

Chapter 3

NITRATE DISSIMILATION IN MARINE AND ESTUARINE SEDIMENTS

R.A. HERBERT

Department of Biological Sciences,
The University, Dundee, Scotland

Introduction

Nitrogen is one of the most important elements required by biological systems and in many marine and estuarine habitats its availability has been frequently implicated as one of the factors limiting productivity [Thomas, 1970; Herbert, 1975; Stewart, 1975]. As a consequence, there is a need to establish the forms in which nitrogen is present in estuarine and marine environments, the routes by which they are transformed, the microorganisms concerned with these processes and the rates at which these processes occur *in situ*. Inorganic nitrogen is found in aquatic environments in a variety of oxidation states ranging from nitrate the most oxidized, to ammonia the most reduced. With the exception of dissolved nitrogen gas, the principal forms of nitrogen occurring in marine and estuarine environments are nitrate, ammonia, nitrite and organic-N. Nitrogen may enter estuarine and inshore coastal waters from a variety of sources including precipitation, agricultural and urban run-off, sewage discharges, sludge dumping and biological nitrogen fixation. Many of these sources are diffuse and these inputs are therefore difficult to determine quantitatively.

In estuarine and inshore coastal waters the concentrations of nitrate, nitrite and ammonia vary considerably from estuary to estuary (Table 1) depending upon the input from urban and agricultural sources. In general, nitrogen levels are highest during winter and early spring and decrease sharply during the early summer as a result of assimilation by phyto- and zoo-plankton. With the exception of heavily polluted estuaries, such as the Tyne and the Mersey, ammonia levels are normally significantly lower than nitrate levels in the water column. Nitrite rarely accumulates in more than trace amounts in the water column of estuaries and inshore waters. In offshore

TABLE 1

Inorganic nitrogen concentrations of several river estuaries in the United Kingdom. Concentrations quoted are mean winter levels and mid estuary location

	expressed as μg N l^{-1}		
Estuary	NO_3-N	NO_2-N	NH_4-N
Tay	760	trace	100
Eden	140	trace	80
Forth	700	trace	300
Tyne	400	trace	760
Humber	470	trace	160
Thames	630	trace	400
Mersey	3,000	90	2,000
Clyde	1,600	trace	90

coastal waters, levels of inorganic nitrogen are significantly lower than those found in estuarine waters (e.g. average winter values for NO_3-N in UK waters ranges from 60 to 100 μg N l^{-1} and for NH_4-N from 10 to 40 μg N l^{-1}).

Whilst nitrate concentrations in the water column are normally higher than the ammonia concentration the reverse has been shown to occur in sediments. For example, Dunn *et al.* [1978] showed that in Kingoodie Bay sediments in the Tay estuary, the mean ammonia concentration (16.1 μg N l^{-1}) was some five times higher than that of nitrate (2.8 μg N l^{-1}). The origin of this ammonia in sediments is normally assumed to be organic nitrogen which is degraded during the process of ammonification. An alternative source of this ammonia, however, is nitrate since a number of facultatively-anaerobic and strictly anaerobic fermentative bacteria have been shown to dissimilate nitrate to produce ammonia at low oxygen tensions [Cole, 1978; Dunn *et al.*, 1978; Herbert *et al.*, 1980]. Sørensen [1978] amended sediments from the Limfjorden in northern Denmark with ^{15}N-labelled nitrate and showed that reduction to ammonia occurred at a rate ranging from 0.12 to 0.75 μmol N cm^{-3} day^{-1}. Similar results have been reported by Koike and Hattori [1978*a*] for sediment samples taken from Mangoku-Ura Bay, Simoda Bay and Tokyo Bay, Japan. These latter workers simultaneously determined nitrate reduction to ammonia and gaseous products (N_2, N_2O) and showed that in the organically richer sediments of Mangoku-Ura Bay 52% of the nitrate dissimilated was reduced to ammonia. In contrast, in the nutritionally poorer Tokyo Bay sediments, only 7% was dissimilated to ammonia and denitrification was the predominant process. The ecological significance of nitrate dissimilation to ammonia is that it conserves nitrogen in a biologically useful form whereas reduction to gaseous products (N_2O and N_2 gas) results in its loss from the ecosystem.

Nitrate Dissimilation in Estuarine and Marine Sediments

We have studied the process of nitrate dissimilation to ammonia at a number of locations on the east coast of Scotland, including a range of river estuaries and a North Sea site. In initial experiments, sediment samples were amended with nitrate and incubated at 15°C in the laboratory for up to 7 days. Data in Table 2 show that sediment samples from all sites produced nitrite following the addition of nitrate compared with unsupplemented controls. Sediment samples from the Don, Tay and site G35 in the North Sea were particularly active and produced considerable quantities of nitrite when supplemented with 10 mM KNO_3. At lower nitrate additions (2 mM KNO_3) ammonia was the major end-product of nitrate dissimilation in several of the sediment-samples (South Esk, Leven and site G35 in the North Sea). From these data it is evident that when nitrate was present in excess the process of nitrate dissimilation proceeded as far as nitrite whereas when nitrate was limiting, reduction continued through to ammonia.

Isolation of Nitrate Dissimilating Bacteria

Two procedures were used to isolate the bacteria involved in the dissimilation of nitrate to nitrite and ammonia in marine and estuarine sediments. The basis of the two methods is outlined in Fig. 1. Direct isolations of the predominant nitrate dissimilating bacteria were made by inoculating a non-selective medium (nutrient agar supplemented with 0.1% w/v KNO_3) with sediment and incubating under an atmosphere of H_2 + CO_2 at 15°C for 7 days. Fifty colonies were then picked off at random from these plates and after checking for purity each isolate was then tested for its ability to grow on nitrate broth under anaerobic conditions. The isolates were then identified to genus level on the basis of a limited number of morphological and biochemical tests according to the scheme of Cowan and Steel [1974]. Data presented in Table 3 show clearly the predominance of *Aeromonas*-*Vibrio* types in all the east-coast river estuaries studied. Isolates identified as belonging to the family Enterobacteriaceae were the second major group present in those sediments rich in fine silts (Don, South Esk, Tay and Leven) whereas pseudomonads predominated in the coarser sediments (Dee and North Esk). From these results it was evident that fermentative bacteria (*Aeromonas*, *Vibrio* and *Klebsiella* species) were predominant and that pseudomonads represented only a small proportion of the total nitrate reducing microflora. Cole and Brown [1980] have reported

TABLE 2

Nitrite and ammonia production after incubation of sediment samples obtained from several east-coast river estuaries and sampling stations in the North Sea

Sampling site	Incubation details	NO_2-N (µg N ml^{-1} sediment)		NH_4-N (µg N ml^{-1} sediment)	
		Time 0	7 days	Time 0	7 days
River Don	No addition	0.10	0.08	0.60	1.17
	2 mM KNO_3	0.08	2.57	0.60	1.12
	10 mM KNO_3	0.07	7.04	0.78	2.15
River South Esk	No addition	0.05	0.56	0.06	0.9
	2 mM KNO_3	0.10	2.97	0.08	5.27
	10 mM KNO_3	0.10	2.36	0.43	3.56
Tay Estuary	No addition	0.00	2.56	0.36	0.54
	2 mM KNO_3	0.15	6.78	0.32	2.78
	10 mM KNO_3	0.15	8.50	0.46	2.82
River Leven	No addition	0.00	0.00	1.58	1.15
	2 mM KNO_3	0.04	0.01	1.75	3.47
	10 mM KNO_3	0.025	0.05	1.73	7.80
North Sea G35	No addition	0.02	0.03	3.22	2.55
	2 mM KNO_3	0.02	0.11	3.43	7.57
	10 mM KNO_3	0.02	22.10	3.20	12.06

TABLE 3

Generic distribution of nitrate respiring bacteria isolated from various river estuaries on the east coast of Scotland

+Genus	River: Don	Dee	North Esk	South Esk	Tay	Leven
Aeromonas/Vibrio	71	54	54	58	48	50
Alcaligenes	9	2	0	6	0	3
Enterobacteriaceae	14	10	10	13	24	28
Pseudomonas	2	27	34	10	18	3
Gram positive bacteria	4	3	2	9	0	10
Others	0	4	0	4	10*	8

* No growth on sub-culture

+ Expressed as a percentage of organisms isolated

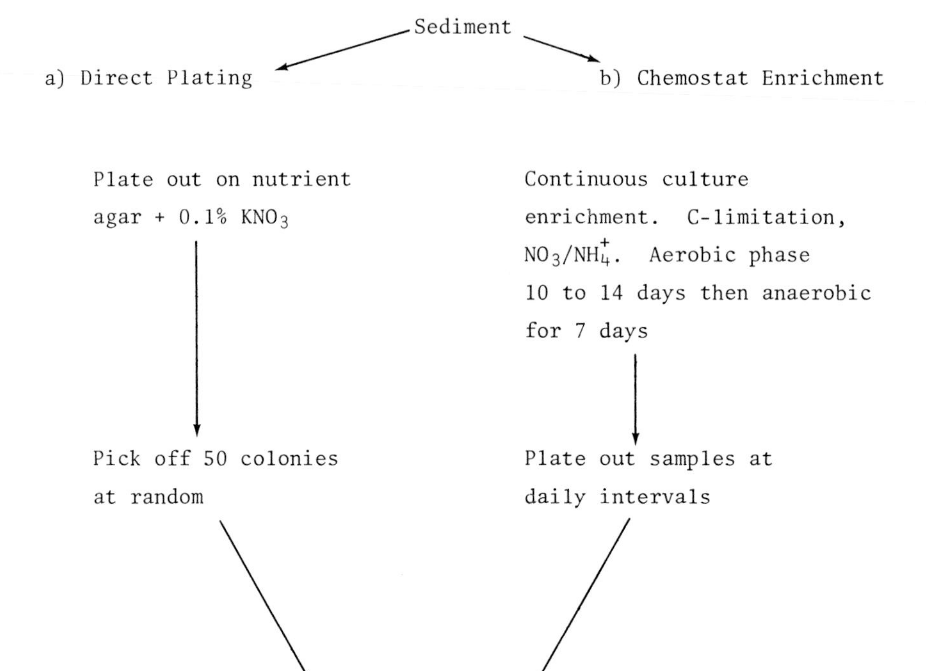

Fig. 1 Scheme for the isolation and identification of nitrate dissimilating bacteria by a) direct plating methods and b) chemostat enrichment procedures.

similar findings for sediment samples obtained from Scottish west-coast sea lochs and sampling sites in the North Atlantic.

The majority of the isolates, identified as *Aeromonas/ Vibrio* and *Klebsiella* sp., failed to produce gas when inoculated into nitrate broth and incubated under anaerobic conditions. Nitrite and ammonia were found to be the principal products of dissimilation. It would thus appear from these data that fermentative bacteria play a significant role in nitrate dissimilation in estuarine and marine sediments, and that the occurrence of oxidative bacteria able to reduce nitrate to gaseous products is relatively rare. Most probable number estimations (MPN) of nitrate reducing bacteria in Kingoodie Bay sediments in the Tay estuary yielded population densities of circa 2 x 10^7 bacteria g dry wt^{-1} sediment during the summer months.

A more sophisticated approach to the isolation of nitrate reducing bacteria from estuarine and marine sediments was to use chemostat enrichments (Fig. 1). Enrichments were carried out in 1 litre capacity chemostats of the type described by Baker [1968]. Carbon limited conditions were employed since in an earlier study Herbert [1975] had reported that Kingoodie Bay sediments in the Tay estuary had low C:N ratios. The medium of Brown *et al.* [1977] modified by the inclusion of either glycerol or acetate as carbon source and containing KNO_3 in addition to NH_4Cl was used. Acetate and glycerol were selected as carbon sources because their metabolism is usually considered to be oxidative. Enrichments were carried out at three salt regimes: freshwater, 17‰ salinity (½ strength sea water) 35‰ salinity (full strength sea water). The inoculum was 25 g sediment suspended in 500 ml of sterile saline of appropriate salinity. The dilution rate was 0.035 h^{-1} and the temperature 15°C. The enrichments were run aerobically for 10 to 14 days, to select against obligate anaerobes, and then anaerobically under a gas phase of high purity nitrogen for a further 7 days. The final bacterial populations therefore consisted of facultatively anaerobic bacteria able to grow anaerobically on glycerol or acetate as carbon source in the presence of nitrate. The bacteria isolated by these procedures were identified according to conventional methods [Cowan and Steel, 1974] and checked for their ability to reduce nitrate under anaerobic conditions. Data presented in Table 4 summarize the results obtained from these enrichment procedures (for more detailed information the reader should consult Dunn *et al.*, 1980). The dominant final populations in the glycerol enrichments, at all salt concentrations, belonged to the family Enterobacteriaceae. More detailed identification procedures showed that these enterobacteria were *Klebsiella* sp. although occasionally *Escherichia coli* was isolated. *Aeromonas/Vibrio* sp. were rarely encountered and it was only in the 17‰ salinity enrichment with glycerol that these organisms formed a significant component of the final population. In the acetate enrichments pseudomonads formed increasingly dominant final populations as the salinity was increased. However, even in the acetate enrichments, enterobacteria were still present in the final populations and in the 17‰ salinity enrichment they become predominant.

As might be anticipated, the two methods of assessing populations of nitrate reducing bacteria produced quite different results. The direct plating method gives information on the identity of the total nitrate reducing population without any indication of which bacteria are most active in this process. Similarly, the continuous culture enrichment system gives no information on the overall populations present but provides extremely useful

TABLE 4

Bacteria isolated from Kingoodie Bay sediments (Tay estuary) by enrichment in continuous culture.
For full experimental details refer to Dunn et al. [*1980*]

Salinity (expressed as ‰)	Carbon Source			
	Glycerol		Acetate	
0	Enterobacteria	77%	*Acinetobacter* sp.	87%
	Pseudomonas sp.	22%	*Pseudomonas* sp.	12%
	Pseudomonas sp.	1%	Enterobacteria	1%
17	Enterobacteria	67%	Enterobacteria	85%
	Aeromonas/Vibrio sp.	33%	*Pseudomonas* sp.	15%
35	Enterobacteria	76%	*Pseudomonas* sp.	82%
	Enterobacteria	17%	Enterobacteria	12%
	Acinetobacter sp.	4%	*Pseudomonas* sp.	6%
	Enterobacteria	3%		

information regarding those bacteria best suited to grow at the expense of nitrate reduction in an anaerobic carbon limited environment. Nevertheless, irrespective of which isolation technique is used, fermentative bacteria appear to be numerically the dominant and most active nitrate reducing organisms present in the estuarine and inshore marine sediments investigated. Recently, we have isolated from Tay estuary sediments strict anaerobes which will respire nitrate with the formation of nitrate and ammonia [Keith and Herbert, unpub. data]. These bacteria have been identified as strains of *Clostridium butyricum*. Other workers [Hasan and Hall, 1975; Caskey and Tiedje, 1980] have also reported the occurrence of clostridia which will reduce nitrate to ammonia with a concomitant increase in cell yield.

Physiology of Nitrate Dissimilation

Detailed physiological studies have been made on a number of nitrate reducing bacteria isolated from Tay estuary sediments by the methods described previously. The object of these studies was to determine the role of nitrate in enabling these bacteria to grow under anaerobic conditions, the products of nitrate dissimilation and how the process was influenced by environmental factors such as oxygen tension and salinity.

Klebsiella K312, isolated from a 17‰ salinity glycerol enrichment, has been the most extensively studied nitrate reducing bacterium [Dunn *et al.*, 1978, 1979; Herbert *et al.*, 1980]. When grown in glycerol limited continuous

TABLE 5

The influence of carbon and nitrogen limitation on the reduction of nitrate by some fermentative bacteria isolated from Kingoodie Bay sediments (Tay estuary)

Bacterial Isolate	Growth conditions		NO_3-N added (mM)	Nitrogen excreted		Nitrate reductase	Nitrite reductase
				NO_2-N	NH_4-N	(nmol min^{-1}	mg^{-1} protein)
Klebsiella K312	N-limited	aerobic	15	N.D.	N.D.	14	N.D.
		anaerobic	15	N.D.	9.4	2058	1314
	C-limited	aerobic	50	0.2	0.1	12	N.D.
		anaerobic	50	5.6	0.8	1814	N.D.
Vibrio V48	N-limited	aerobic	10	N.D.	N.D.	16	N.D.
		anaerobic	10	N.D.	0.8	2000	373
	C-limited	aerobic	20	0.39	N.D.	200	N.D.
		anaerobic	20	4.25	N.D.	625	N.D.
Clostridium butyricum SS-6	N-limited	aerobic	3.5	N.G.	N.G.	N.G.	N.G.
		anaerobic	3.5	N.D.	1.2	123	56
	C-limited	aerobic	10	N.G.	N.G.	N.G.	N.G.
		anaerobic	10	3.0	0.1	98	N.D.

N.D. = not detected

N.G. = no growth

culture under anaerobic conditions with nitrate, the end-product of nitrate dissimilation was principally nitrite which was excreted into the growth medium (Table 5). Nitrate dissimilation under anaerobic conditions was associated with an active membrane bound nitrate reductase. Under nitrate limitation, cultures of *Klebsiella* K312 excreted ammonia as the end-product of nitrate dissimilation in the absence of oxygen and both a particulate nitrate reductase and a soluble nitrite reductase were synthesized. The excreted ammonia, under anaerobic conditions, accounted for 62% of the nitrate present in the medium reservoir. Ammonia did not repress the synthesis of either nitrate reductase or nitrite reductase unless nitrate was omitted from the medium. Glycerol limited cultures of *Klebsiella* K312 grown on ammonia under anaerobic conditions did not synthesize either a nitrate or nitrite reductase and the absence of an alternative electron acceptor to molecular oxygen was reflected in the sharp decrease in cell population density (Table 6).

TABLE 6

Influence of carbon and nitrogen limitation under aerobic and anaerobic conditions on cell population densities of Klebsiella *K312 grown in continuous culture*

Growth Conditions	Population density (ml culture)$^{-1}$		Population % remaining
C-limited/NH_4^+	aerobic	8.9×10^9	6%
	anaerobic	5.4×10^8	
C-limited/NO_3^-	aerobic	2.3×10^9	34%
	anaerobic	8×10^8	
N-limited/NO_3^-	aerobic	4.9×10^9	26%
	anaerobic	1.3×10^9	

Upon switching from aerobic to anaerobic conditions with ammonia as nitrogen source, the population density decreased to 6%. In contrast, when nitrate was substituted for ammonia, the population densities whilst still declining under anaerobic conditions were significantly higher (Table 6), reflecting the role of nitrate as an alternative electron acceptor to molecular oxygen.

Data presented in Tables 7 and 8 show the effect of salinity on nitrate dissimilation by *Klebsiella* K312. The overall effect appears to be a decrease in total population density compared with cultures grown in the absence of salt. The products of nitrate dissimilation

TABLE 7

Influence of salinity on nitrate dissimilation by Klebsiella *K312 grown in continuous culture. Salt concentration equivalent to a salinity of 17‰*

Growth conditions		NO_3-N added (mM)	Nitrogen excreted		Nitrate reductase	Nitrite reductase
			NO_2-N	NH_4-N	(nmol min^{-1} mg^{-1} protein)	
C-limited/NH_4-N	aerobic	0	-*	-*	16	N.D.
	anaerobic		-	-	35	N.D.
N-limited/NO_3-N	aerobic	15	0.1	0.1	36	21
	anaerobic		0.2	5.6	2317	976
C-limited/NO_3-N	aerobic	50	0.1	0.1	13	22
	anaerobic		2.9	0.7	1042	N.D.

* No data

N.D. not detected

TABLE 8

Influence of salinity on nitrate dissimilation and cell population densities of Klebsiella *K312 grown in continuous culture. Salt concentration equivalent to a salinity of 17‰*

Growth conditions		Population density (ml culture)$^{-1}$	% population remaining
C-limited/NH_4^+	aerobic	1.6×10^9	4
	anaerobic	6.7×10^7	
C-limited/NO_3^-	aerobic	2×10^8	40
	anaerobic	0.8×10^8	
N-limited/NO_3^-	aerobic	7.2×10^8	30.5
	anaerobic	2.2×10^8	

are influenced by nitrate availability and not salinity, although quantitatively the amounts of nitrite and ammonia produced were significantly less than those produced in the absence of salt. The decreased cell populations observed when the salinity was increased to 17‰ may reflect the increased utilization of the carbon source for osmoregulatory purposes.

While it has been well established that nitrate dissimilation occurs maximally in the absence of molecular oxygen, there have been relatively few detailed studies on the influence of oxygen tension on this process and its effects on nitrate reductase and nitrite reductase. We have made a detailed study of the effects of oxygen tension on nitrate dissimilation by *Klebsiella* K312 and data presented in Table 9 summarize the results of these investigations [Dunn *et al.*, 1979]. From the data presented in Table 9 it is evident that in aerobic cultures (150 mm Hg dissolved oxygen tension) grown under carbon or nitrate-limited conditions only trace quantities of nitrite and ammonia were produced in the spent medium. These cultures contained low levels of nitrate reductase and no detectable nitrite reductase. Decreasing the dissolved oxygen tension of the culture to 15 mm Hg (circa 10% air saturation) resulted in the synthesis of a particular nitrate reductase and in nitrate limited cultures a soluble nitrite reductase. In nitrate limited cultures ammonia but not nitrite was excreted into the medium at this oxygen tension, whereas carbon limited cultures accumulated neither nitrite nor ammonia.

TABLE 9

Influence of oxygen tension on nitrate dissimilation by Klebsiella *K312 grown in continuous culture*

Growth conditions	Oxygen tension (mm Hg)	NO_3-N added (mM)	Nitrogen excreted		Nitrate reductase	Nitrite reductase
			NO_2-N	NH_4-N	(nmol min^{-1} mg^{-1} protein)	
N-limited	150	15	-	-	14	N.D.
	15		-	2.3	195	113
	9		0.1	2.1	445	58
	N_2 flow		-	9.4	2058	1314
C-limited	150	50	0.2	0.1	12	N.D.
	15		0.2	0.1	370	N.D.
	8		0.8	0.2	704	N.D.
	N_2 flow		5.6	0.8	1814	N.D.

N.D. not detected

Decreasing the oxygen tension progressively to zero in nitrate limited cultures led to the excretion of increased quantities of ammonia until under fully anaerobic conditions (N_2 gas flow) 62% of the nitrate in the medium was utilized this way. Nitrite never accumulated in more than trace quantities in nitrate limited culture at any oxygen tension. As the oxygen tension was decreased the activities of both nitrate reductase and nitrite reductase increased markedly reaching a maximum under fully anaerobic conditions. In marked contrast, carbon limited cultures excreted only small quantities of ammonia while nitrite was the principal product of dissimilation. Under fully anaerobic conditions nitrite excretion accounted for 12% of the nitrate input whilst only 2% of the nitrate utilized was excreted as ammonia. Nitrite reductase was never observed in carbon limited cultures of *Klebsiella* K312. It is apparent from these data that at oxygen tensions of 15 mm Hg and below, this bacterium can utilize molecular oxygen, nitrate and nitrite simultaneously as terminal electron acceptors. It is something of a paradox that under nitrate-limitation the excreted ammonia (Table 9) was not assimilated and under such conditions these must be considered to be electron acceptor-limited cultures. Similar data have been reported by Cole [1978] for *E. coli* although in this organism ammonia was excreted irrespective of whether cultures were grown under carbon or nitrogen limitation. When nitrite (5 mM) was substituted for nitrate, *Klebsiella* K312 grew only slowly under aerobic conditions possibly because of toxic effects. Under anaerobic conditions, however, nitrite was utilized, both nitrate and nitrite reductases were synthesized and approximately 30% of the nitrite-nitrogen utilized was excreted as ammonia. Cole [1978] reported that the growth yield of *E. coli* grown on nitrite was similar to that during growth on ammonia suggesting that nitrite reduction was essentially an electron sink mechanism with no oxidative phosphorylation involved. It may well be that this was also the situation with *Klebsiella* K312 and this would explain why this bacterium preferentially reduced nitrate. These data for *Klebsiella* K312 indicate that this bacterium can dissimilate nitrate under microaerobic conditions which contrasts with the data of Justin and Kelly [1978] for *Thiobacillus denitrificans*. Whilst *T. denitrificans* synthesized both nitrate and nitrite reductases at low oxygen tensions, nitrate reduction in this bacterium only occurred anaerobically.

In contrast to the data on *Klebsiella* K312 there is less information on the *Aeromonas/Vibrio* sp. which are the numerically dominant nitrate reducing bacteria present in the estuarine and marine sediments investigated. Initial studies with a *Vibrio* sp. designated V25 suggested that this bacterium played only a limited role in the process of nitrate reduction since it was unable to reduce nitrate beyond nitrite even under nitrate limiting conditions [Herbert *et al.*, 1980]. Carbon limited cultures of *Vibrio* V25 achieved bacterial population densities approximately 33% of that obtained aerobically, whereas under nitrate limitation cell populations decreased to 6% of those observed aerobically. A systematic search for *Vibrio/Aeromonas* sp. which were more active in nitrate reduction has yielded a number of isolates which will dissimilate nitrate through to ammonia under anaerobic conditions. Data presented in Table 5 show results obtained for *Vibrio* V48 isolated from Tay estuary sediments by conventional plate count methods. Under nitrate limited conditions in the absence of oxygen this bacterium dissimilated nitrate through to ammonia without the accumulation of nitrite as an intermediate. This reduction of nitrate was associated with the synthesis of active nitrate and nitrite reductases. Under carbon limitation, nitrite accumulates as the end-product of nitrate dissimilation and only a particulate nitrate reductase was synthesized. When grown anaerobically on nitrate under carbon- or nitrate-limitation the population densities of *Vibrio* V48 decreased in a manner analogous to that described for *Klebsiella* K312 (Table 6). Similar results have also been obtained by Macfarlane and Herbert [unpub. data] for type strains of *Vibrio parahaemolyticus* (NCMB 1903 and NCMB 2047). It appears therefore, that not all the *Vibrio* sp. are restricted in their ability to dissimilate nitrate under anaerobic conditions.

Nitrate Dissimilation by Strict Anaerobes

A number of workers have reported that clostridia and in particular *Clostridium perfringens* can dissimilate nitrate to ammonia with a concomitant increase in cell yield compared with fermentative growth [Woods, 1938; Hasan and Hall, 1975; Chiba and Ishimoto, 1977; Caskey and Tiedje, 1980]. Saccharolytic clostridia isolated from sediments in the Tay estuary using chemostat enrichment methods were subsequently identified as strains of *Clostridium butyricum*. When *Cl. butyricum* SS-6 was grown under carbon-limitation with nitrate, significant quantities of nitrite were excreted into the growth medium (30% of the nitrate utilized) and this was associated with the synthesis of a soluble nitrate reductase (Table 5). Under nitrate limitation, ammonia was the sole product of nitrate reduction and both a nitrate and a nitrite reductase was synthesized. Corresponding cultures grown on ammonia did not synthesize either a nitrate or a nitrite reductase and cell yields were 20% lower than those of nitrate grown cultures. In contrast to the dissimilatory membrane bound nitrate reductases of *Pseudomonas denitrificans* [Chang and Morris, 1962], *Klebsiella* spp. [Van't Riet *et al.*, Dunn *et al.*, 1978], *E. coli* [Showe and De Moss, 1968] the enzyme in clostridia is found in the soluble fraction [Chiba and Ishimoto, 1973, 1977].

Pulsing a nitrate grown culture of *Cl. butyricum* with 3.5 mM ammonia did not reduce nitrate and nitrite reductase activities significantly and this is consistent with their dissimilatory role in this bacterium. Conversely, pulsing an ammonia grown culture with 3.5 mM nitrate induced both a nitrate and nitrite reductase.

Discussion

Data presented in this review show that a number of facultatively-anaerobic and strictly anaerobic bacteria obtained from inshore marine and estuarine sediments are capable of dissimilating nitrate to either nitrite or ammonia, depending upon nitrate availability. It is evident that energetically the process of nitrate respiration is superior to that of fermentation and this is shown clearly by the increased population densities and cell yield of these bacteria when grown in the presence of nitrate under anaerobic conditions. *Klebsiella* K312 and allied types are physiologically the most versatile of the isolates so far studied. The role of the *Aeromonas/ Vibrio* isolates remains unclear. Numerically they are dominant as determined by the direct plate count method yet they never predominate in any of the chemostat enrichments. Initially Herbert *et al.* [1980] considered that the *Vibrio* sp. were unable to compete effectively with the enterobacteria since they were unable to reduce nitrate beyond nitrite and thus would become electron acceptor limited under conditions of nitrate limitation. This view-point must however be re-evaluated in the light of data obtained with *Vibrio* V48 and type strains of *Vibrio parahaemolyticus* which show unequivocally that certain species can dissimilate nitrate to ammonia under nitrate limiting conditions. The role of the saccharolytic clostridia in nitrate reduction to ammonia remains equally unclear since no detailed studies of population densities in estuarine and marine sediments have been made. However, they may have a role more significant than just the reduction of nitrate to nitrite and ammonia. Studies in this laboratory have shown that when *Clostridium butyricum* SS6 was grown under ammonia limitation the principal fermentation product was butyric acid whereas when dissimilating, nitrate acetate accumulated in the spent media. Similar findings have been reported previously by Ishimoto *et al.* [1974] for *Cl. perfringens*. The ability of the clostridia to accumulate different end-products of cellulose catabolism depending upon the availability of nitrate may have considerable significance in terms of energy flow within marine and estuarine sediments.

Insufficient studies have been made to determine the relative roles of nitrate dissimilation to ammonia compared with denitrification to gaseous products. Sørensen [1978] in his study of nitrate reduction in the sediments of the Limfjorden, Northern Denmark concluded that nitrate dissimilation to ammonia was probably equally important as denitrification in the turnover of nitrate in marine sediments. In a more extensive study Koike and Hattori [1978*a,b*] showed that in sediments of Simoda Bay and Tokyo Bay, denitrification was the major route for the anaerobic reduction of nitrate whereas in the organically richer Mangoku-Ura Bay sediments dissimilation to ammonia was the dominant process. These workers were also able to demonstrate that the two processes could proceed simultaneously within anaerobic inshore marine sediments.

The process of the anaerobic dissimilation of nitrate to ammonia is the reverse of nitrification, the aerobic oxidation of ammonia to nitrate by nitrifying bacteria. Since sharp discontinuities in oxygen tension occur in marine and estuarine sediments over extremely small vertical distances or within microniches it is probable that *in situ* both processes occur simultaneously. Experimental evidence to support this proposition has been provided by Koike and Hattori [1978*b*] who demonstrated using ^{15}N-labelled methods the simultaneous processes of nitrification and nitrate reduction to ammonia in sediments of Odawa Bay, Japan. These authors concluded that the conditions prevailing in these coarse sandy sediments permitted nitrifying bacteria to grow whilst organic detritus from adjacent *Zostera* beds provided sufficient organic matter to maintain reduced micro-environments within otherwise aerobic sediments. It is evident from the results presented in this chapter that with the *Klebsiella* sp. nitrate reduction proceeds under micro-aerobic conditions and there is therefore no requirement for totally anoxic conditions for nitrate reduction to occur. The operation of such an internal cycle is clearly energetically advantageous to both groups of bacteria and more detailed studies are now required to substantiate the work of Koike and Hattori [1978*a,b*].

Conclusions

Fermentative bacteria isolated from a number of east-coast river estuaries and inshore sampling sites in the North Sea appear to be numerically and potentially the most active nitrate reducing bacteria within these sediments. Members of the family Enterobacteriaceae and in particular *Klebsiella* sp. appear to be the most versatile of the nitrate dissimilating bacteria isolated. Depending upon the availability of nitrate the end-products of nitrate dissimilation may be either nitrite or ammonia. The dissimilatory reduction of nitrate in the fermentative bacteria is associated with the synthesis of a membrane bound nitrate reductase and under certain growth conditions a nitrite reductase whilst in the clostridia examined the dissimilatory nitrate reductase was in the soluble fraction. Nitrate reduction enables these bacteria to maintain high cell population densities under anaerobic conditions due to its role as an electron acceptor.

The relationship between the anaerobic dissimilation of nitrate to ammonia, ammonification and nitrification in estuarine and marine sediments are poorly understood and are clearly well worthy of detailed study.

Acknowledgement

The author is grateful to the Natural Environmental Research Council for financial support for this work through Grants GR3/2729 and GR3/4208.

References

Baker, K. (1968). Low cost continuous culture. *Laboratory Practice* **17**, 817-824.

Brown, C.M., Ellwood, D.C. and Hunter, J. (1977). Growth of bacteria at surfaces: Influence of nutrient limitation. *FEMS Letters* **1**, 163-166.

Caskey, W.H. and Tiedje, J.M. (1980). The reduction of nitrate to ammonium by a *Clostridium* sp. isolated from soil. *Journal of General Microbiology* **119**, 217-223.

Chang, J.P. and Morris, J.G. (1962). Studies on the utilization of nitrate by *Micrococcus denitrificans*. *Journal of General Microbiology* **29**, 301-310.

Chiba, S. and Ishimoto, M. (1973). Ferredoxin linked nitrate reductase from *Clostridium perfringens*. *Journal of Biochemistry, Tokyo* **73**, 1315-1318.

Chiba, S. and Ishimoto, M. (1977). Studies on nitrate reductase of *Clostridium perfringens*. I. Purification, some properties and effect of tungstate on its formation. *Journal of Biochemistry* **82**, 1663-1671.

Cole, J.A. (1978). The rapid accumulation of large quantities of ammonia during nitrate reduction by *E. coli*. *FEMS Microbiology Letters* **4**, 327-329.

Cole, J.A. and Brown, C.M. (1980). Nitrate reduction to ammonia by fermentative bacteria: short circuit of the biological nitrogen cycle. *FEMS Microbiology Letters* **7**, 65-72.

Cowan, S.T. and Steel, K.J. (1974). Manual for the Identification of Medical Bacteria. Cambridge University Press.

Dunn, G.M., Herbert, R.A. and Brown, C.M. (1978). Physiology of denitrifying bacteria from tidal mudflats in the River Tay. In "Physiology and Behaviour of Marine Organisms" (Eds. D.S. McLusky and A.J. Berry), pp.135-140. Pergamon Press, Oxford.

Dunn, G.M., Herbert, R.A. and Brown, C.M. (1979). Influence of oxygen tension on nitrate reduction by a *Klebsiella* sp. growing in chemostat culture. *Journal of General Microbiology* **112**, 379-383.

Dunn, G.M., Wardell, J.N., Herbert, R.A. and Brown, C.M. (1980). Enrichment, enumeration and characterization of nitrate reducing bacteria present in sediments of the River Tay. *Proceedings of the Royal Society of Edinburgh 78B*, 47-56.

Hasan, S.M. and Hall, J.B. (1975). The physiological function of nitrate reduction in *Clostridium perfringens*. *Journal of General Microbiology* **87**, 120-128.

Herbert, R.A. (1975). A preliminary investigation of the effects of salinity on the bacterial flora of the Tay estuary. *Proceedings of the Royal Society of Edinburgh 78B*, 137-144.

Herbert, R.A., Dunn, G.M. and Brown, C.M. (1980). The physiology of nitrate dissimilatory bacteria from the Tay estuary. *Proceedings of the Royal Society of Edinburgh 78B*, 79-87.

Ishimoto, M., Umeyama, M. and Chiba, S. (1974). Alteration of fermentation products from butyrate to acetate by nitrate reduction in *Clostridium perfringens*. *Zeitschrift für Allgemeine Mikrobiologie* **14** (2), 115-121.

Justin, P. and Kelly, D.P. (1978). Metabolic changes in *Thiobacillus denitrificans* accompanying the transition from aerobic to anaerobic growth in continuous culture. *Journal of General Microbiology* **107**, 131-137.

Koike, I. and Hattori, A. (1978*a*). Denitrification and ammonia formation in anaerobic coastal sediments. *Applied and Environmental Microbiology* **35**, 278-282.

Koike, I. and Hattori, A. (1978*b*). Simultaneous determinations of nitrification and nitrate reduction in coastal sediments by a ^{15}N dilution technique. *Applied and Environmental Microbiology* **35** (5), 853-857.

Showe, M.K. and De Moss, J.A. (1968). Localization and regulation of nitrate reductase in *E. coli*. *Journal of Bacteriology* **95**, 1305-1313.

Sørensen, J. (1978). Capacity for denitrification and reduction of nitrate to ammonia in a coastal marine sediment. *Applied and Environmental Microbiology* **35** (2), 301-305.

Stewart, W.D.P. (1975). Biological cycling of nitrogen in untertidal and supralittoral marine environments. *Proceedings 9th European Marine Biology Symposium* (Ed. H. Barnes), pp.637-660. Aberdeen University Press, Aberdeen.

Thomas, W.H. (1970). Effect of ammonium and nitrate concentration on chlorophyll increases in natural tropical Pacific phytoplankton populations. *Limnology and Oceanography* **15**, 386-394.

Van't Riet, J., Stouthamer, A.H. and Planta, R.J. (1968). Regulation of nitrate assimilation and nitrate respiration in *Aerobacter aerogenes*. *Journal of Bacteriology* **96**, 1455-1464.

Woods, D.D. (1938). Reduction of nitrate to ammonia by *Clostridium welchii*. *Biochemical Journal* **32**, 2000-2012.

Chapter 4

THE CYCLING OF SULPHUR IN MARINE AND FRESHWATER SEDIMENTS

D.B. NEDWELL

Department of Biology, University of Essex, Wivenhoe Park, Colchester, UK

Introduction

Sulphur can exist in a number of oxidation states [Thorstensen, 1970] from the most oxidized-sulphate, to the most reduced-sulphide. In aerobic oxidized environments sulphate is the thermodynamically stable state but in reduced anaerobic environments sulphide is the most stable. However, the reduction of sulphate to sulphide in reduced environments does not occur spontaneously at the physical conditions existing in the biosphere, but requires the catalytic mediation of biological activity. (Trudinger [1969] has reviewed the biological transformations of sulphur). The biological reduction of sulphate can take place in two ways.

Assimilatory reduction. Sulphur is required for biosynthesis of cytoplasmic components, principally amino-acids and nucleic acids. It is present in cytoplasm as sulphydryl (HS^-) groups in which sulphur is at the most reduced oxidation state, corresponding to sulphide. If sulphate is the available source of sulphur it must therefore be reduced during its assimilation before it is incorporated in biosynthesis. However, the proportion of sulphur in cytoplasm is low, approximately 1% of dry weight, compared to about 50% for carbon. There is a comparatively low requirement for sulphur for biosynthesis, therefore, and sulphur is rarely a growth-limiting nutrient in aquatic environments. Consequently, only a relatively small amount of sulphur is liberated on the breakdown and mineralization of organic detritus.

Dissimilatory reduction. Under anaerobic conditions a specialized group of anaerobic bacteria, the sulphate-reducing bacteria [Postgate, 1979], can use sulphate as the terminal electron acceptor in their respiration.

These bacteria are generally organotrophic, oxidizing low molecular weight organic compounds and simultaneously reducing sulphate to sulphide. Their metabolism can be represented as:

$$2CH_2O + SO_4^{2-} \rightarrow H_2S + 2HCO_3^- \quad \text{[Berner, 1974]}$$

Therefore the oxidation of each two moles of carbon requires the reduction of a mole of sulphate (i.e. C:S = 3:4, by weight). Such reduction is termed "dissimilatory" sulphate reduction. In the marine environment there is an abundant supply of sulphate (2.7 g kg^{-1} seawater at a salinity of 35‰), and where sulphate reducers are active in the marine environment dissimilatory reduction of sulphate far outstrips assimilatory reduction. This may not be the case in freshwater environments where there is a smaller quantity of sulphate present.

Sulphate Dissimilation

Sulphate reduction requires anaerobic, reduced conditions before sulphate-reducing bacteria are active. Therefore it is only in the bottom sediments of lakes, rivers, estuaries and seas that the complete sulphur cycle is seen, as the water column is usually aerobic and oxidized. Exceptions to this may occur where stratification in the water column reduces the transport of oxygen down into the hypolimnion. Rapid respiration of oxygen by aerobic microbes in the bottom sediments may then exceed its rate of replacement from the epilimnion, with the result that the hypolimnion tends to become deoxygenated, the reduced anaerobic zone extending upward from the bottom sediment and into the water column. Sulphate reduction can then proceed in the anaerobic water column itself, as is found in some deep oceanic trenches [Tuttle and Jannasch, 1973], the Black Sea [Sorokin, 1962], a number of deep fjords [see Deuser, 1975], meromictic lakes [Sorokin, 1970], and some eutrophic lakes during Summer stratification [Jones, 1976; Jones, this volume].

A "typical" sediment may be thought of as having a surface aerobic layer of variable depth. Within this layer oxygen removed by aerobic bacterial respiration is replaced by further oxygen transported down from the sediment/water interface. Such transport may partly be the result of physical diffusion, but overturn and mixing of the surface layer of sediment by invertebrate infauna (bioturbation) and irrigation of the surface layer by respiratory water currents of invertebrate animals may also contribute substantially to transport. However, as depth below the sediment surface increases, the dissolved oxygen concentration diminishes as respiratory removal by microorganisms exceeds replacement. The extent of the surface aerobic zone may vary markedly from a few millimetres to many metres of depth, depending upon the factors which influence oxygen transport and concentration. These factors include the particle size of the sediment, the organic content because of its influence upon microbial respiration rates, temperature, tidal flushing, and bioturbation.

With increased depth, therefore, the sediment becomes increasingly reduced, and anaerobic bacterial metabolism may ensue. Generally, a progression of bacterial metabolism with depth is hypothesized [Mechalas, 1974; Billen, this volume] with a zone of nitrate reduction underlying the aerobic zone, followed by a zone of sulphate reduction, below which is a zone of carbon dioxide reduction (methanogenesis). However, the extent, occurrence and importance of each of these metabolic zones may vary. In freshwater sediment, nitrate reduction may be important and sulphate reduction limited by low sulphate concentrations. For example, Jones and Simon [1980] concluded that in the profundal zone of Blelham Tarn, UK, aerobic respiration accounted for about 42% of organic carbon oxidation, denitrification for 17%, sulphate reduction for only 2%, and methanogenesis for 25%. In contrast, in marine sediments sulphate reduction may be a major mechanism of carbon oxidation, but nitrate reduction comparatively unimportant and limited by low nitrate concentrations. Sørensen *et al.* [1979] calculated that in two Danish coastal marine sediments aerobic respiration accounted for 110 mmols carbon m^{-2} day^{-1}, sulphate reduction 30.2 mmols C m^{-2} day^{-1}, and nitrate reduction only 0.05 mmols C m^{-2} day^{-1}. Jørgensen [1977*c*] estimated that 53% of the organic carbon mineralization in surface sediment of the Danish Limfjord was due to the activity of the sulphate reducing bacteria. Thus, in marine ecosystems sulphate reduction is of great importance.

Sulphate reduction cannot proceed until the sedimentary environment is anaerobic, and the redox potential less than about -100 mv. Active sulphate reduction has been detected using ^{35}S sulphate [Jørgensen, 1977*a*] within an apparently oxidized layer of marine sediment, but this has been attributed to the presence of anaerobic, reduced microsites within the generally aerobic layer. Jørgensen suggested that the centre of organically-rich molluscan faecal pellets might provide such reduced microenvironments, the rapid oxygen consumption by aerobic organotrophic bacteria on the outside of the pellets preventing oxygen penetrating to the centres, where sulphate reduction could therefore occur. The minimum diameter of such microsites which was necessary to ensure a deoxygenated centre varied depending upon the dissolved oxygen concentration in the sediment pore water, and upon the rate of bacterial respiration of oxygen. Using reasonable estimates of these two variables, the calculated diameters of such microsites fell within the size limits of the faecal pellets, supporting the hypothesis of oxygen-depleted microenvironments within such pellets.

Factors Influencing Rate of Sulphate Reduction

Once a suitable physical/chemical environment is present which permits sulphate reduction, the rate of activity is influenced by a number of environmental variables. Generally, measured rates of sulphate reduction exhibit a maximum near the top of the anaerobic zone where concentrations of electron donors and electron acceptor (both derived by transport from the sediment surface) will be greatest. The rate of sulphate reduction then decreases more or less rapidly with increased depth (Fig. 1). Following this trend, counts of sulphate-reducing bacteria also tend to decrease with increased depth (Fig. 1) but the numbers of bacteria counted do not usually correlate with the measured rate of sulphate reduction [e.g. Jørgensen, 1977*c*; Abdollahi and Nedwell, 1979] suggesting that the rate is not usually limited by bacterial numbers. However, this assumes that the technique and medium used to obtain the count adequately reflected the active population of sulphate reducers responsible for the measured rate of sulphate reduction. This may not be true, especially as lactate has almost invariably been used as the substrate in the medium used to count sulphate-reducing bacteria. Groups of sulphate-reducing bacteria metabolizing other substrates are now known (see p. 82) which may not be recovered on a conventional lactate medium, and the lack of correlation might reflect the inadequacy of the media used.

The rate of sulphate reduction will be influenced by the availability of both electron donors and electron acceptor. The effect of sulphate concentration upon the rate of sulphate reduction in marine sediments has been variously described as a zero order reaction in the surface layers where sulphate concentrations approach those of seawater [Rees, 1973; Goldhaber and Kaplan, 1974]; or as a first order reaction with sulphate reduction becoming sulphate-limited below about 5 to 10 mM sulphate [Nakai and Jensen, 1964]. In the upper part of the anaerobic zones of most marine and intertidal sediments, where the rate of sulphate reduction is greatest, the concentration of sulphate in the porewater is often not substantially diminished compared to seawater [for example, see Nedwell and Floodgate, 1972*b*], and is unlikely to be at rate-limiting concentrations. This is because of the comparatively rapid exchange of sulphate from seawater into the upper layers of sediment. In freshwater sediments [Winfrey and Zeikus, 1977; Molongoski and Klug, 1980], and at great depth in marine sediments when influx of further sulphate is negligible because of the diffusion distance, sulphate may be diminished below the concentration at which it becomes limiting to sulphate reduction rates i.e. rates become first order with respect to sulphate. This situation is compatible with Michaelis-Menten saturation kinetics where rates of reduction will be

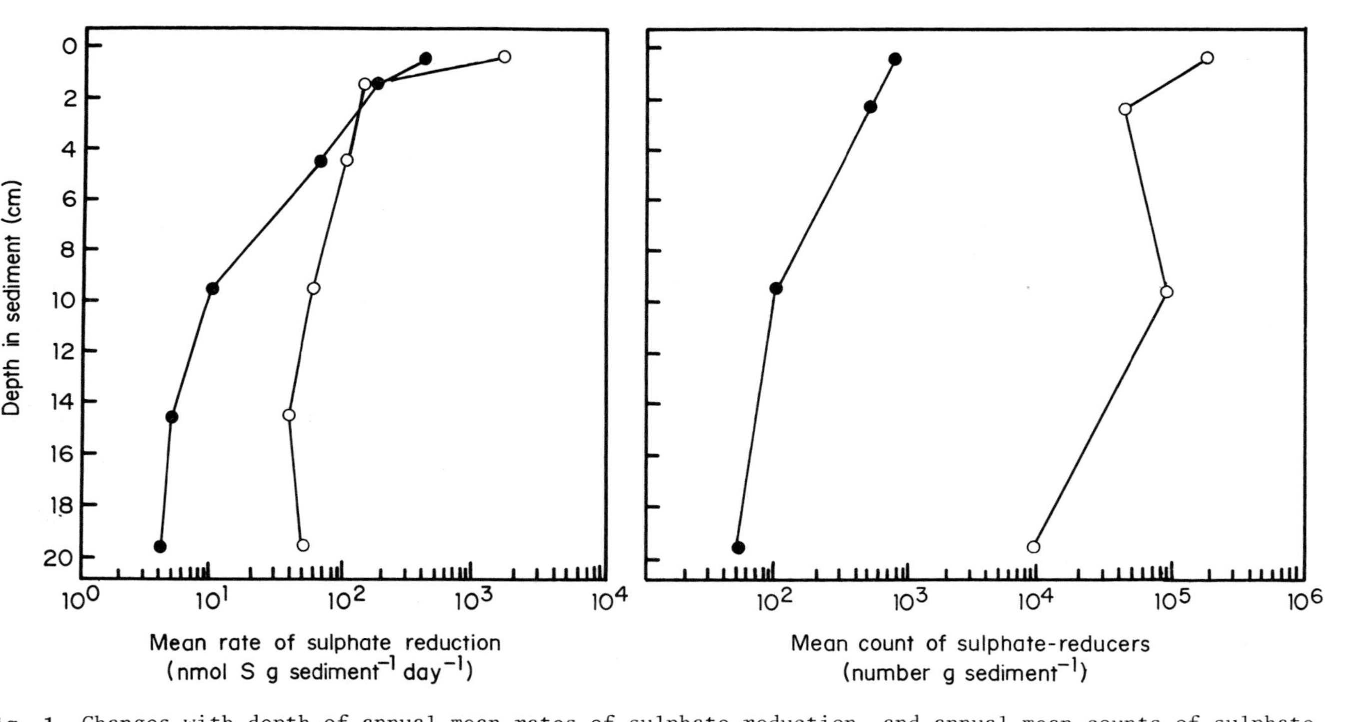

Fig. 1 Changes with depth of annual mean rates of sulphate reduction, and annual mean counts of sulphate reducing bacteria, at two sites in a salt marsh. Open circles - sediment from a creek bottom; closed circles - sediment from a pan. [Redrawn from Nedwell and Abram, 1978].

essentially zero order at high sulphate concentrations and first order at low sulphate concentrations. Indeed Ramm and Bella [1974] have described the sulphate reduction rates in laboratory sediment microcosms in terms of a Michaelis-Menten equation involving both sulphate concentrations and the concentration of Soluble Organic Carbon (SOC) which is presumably an indirect measure of the organic electron donor concentration.

$$\frac{dSO_4}{dt} = R_{max} \left(\frac{SO_4}{K_{SO_4} + SO_4}\right) \left(\frac{SOC}{K_{SOC} + SOC}\right)$$

Where R_{max} is the maximum rate of sulphate reduction;
SO_4 and SOC the concentrations of sulphate and SOC respectively;
K_{SO_4} and K_{SOC} the half saturation constants with respect to sulphate and SOC.

This equation emphasizes the dependence of the rate of reduction upon both electron donor and electron acceptor, although where sulphate is in excess and non-limiting, as in surface marine sediments, the sulphate term will become essentially constant and the rate dependent only upon the electron donor concentration. In contrast, in freshwater sediments it may be sulphate which is limiting to the rate of reduction.

Sulphate dissimilation is also strongly influenced by temperature and marked seasonal changes in sulphate reduction may be seen in coastal sediments (Fig. 2). In an examination of sulphate reduction rates in salt-marsh sediment [Nedwell and Abram, 1979] it was found that temperature change had the greatest influence upon the measured rate, followed by electron donor (lactate) concentration. Sulphate concentration had no effect upon the rate unless the sulphate concentration was diminished well below that normally present in these surface sediments. The effect of temperature upon sulphate reduction can be described by the Arrhenius equation:

$$v = A\ e^{-E/RT}$$

Where A is a constant
E is a "temperature characteristic"
R is the Gas Constant
T is the temperature (°K).

In saltmarsh sediment sulphate reduction had a temperature characteristic of 20.4 kcals mole^{-1}, equivalent to a Q_{10} of 3.5 [Abdollahi and Nedwell, 1979]. Other reported values of Q_{10} for sulphate reduction are

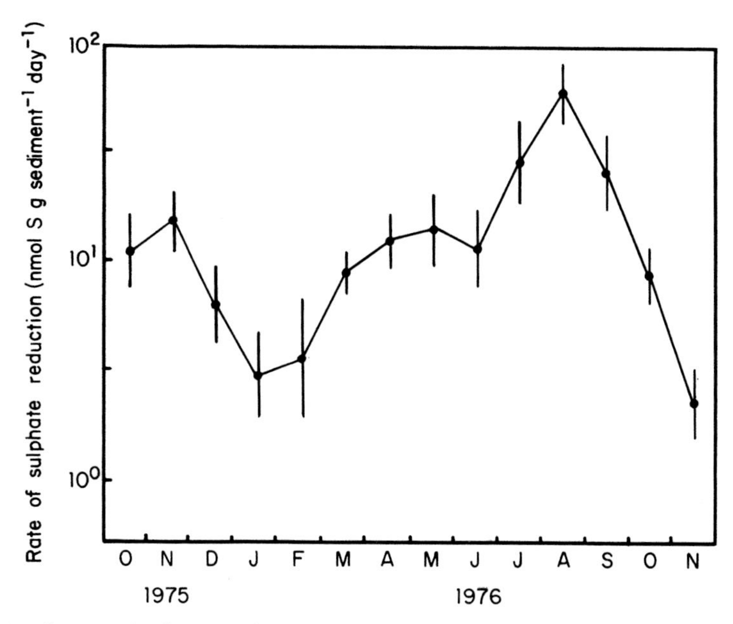

Fig. 2 Seasonal changes in the rate of sulphate reduction in surface sediment from a saltmarsh creek (95% confidence limits shown). [From Abdollahi and Nedwell, 1979].

3.4 in marine sediment from the Danish Limfjord [Jørgensen, 1977*c*], and 3.9 in sediment from the Dutch Wadden Sea [Vosjan, 1975]. At least in the upper layers of coastal marine sediments temperature seems to be the predominant regulator of sulphate reduction with changes in electron donor concentration having less importance. Changes in sulphate concentration have little effect, although this variable is probably much more important in freshwater sediments where sulphate concentrations are always low and where sulphate may entirely disappear during periods of high microbial activity. Again, temperature variation will presumably be of little importance in deep ocean sediments where the environmental temperature is constantly low throughout the year.

As described above, the availability of electron donors will play an important part in regulating the rate of sulphate reduction in marine sediments, and in general terms there is a correlation between the organic content of a sediment and the magnitude of sulphate reduction. Berner [1970] found an approximately linear relationship between the organic content of the upper layer of marine sediments and the amounts of pyrite (FeS_2), which is derived from sulphate reduction. Table 1 summarizes much of the data on rates of sulphate reduction reported in the literature for the surface layers of a variety of marine and intertidal sediments which are rich in sulphate

TABLE 1

Some rates of sulphate reduction reported in the literature for surface layers of marine sediments

Site	Sediment layer	Rate mmol l^{-1} day^{-1}	Source
Intertidal			
Peat from US saltmarsh during summer		0.25 - 6.0	Howarth and Teal [1979]
Saltmarsh pan, UK (seasonal limits)	0-10 cm	0.0017 - 0.19	Nedwell and Abram [1978]
Saltmarsh creek, UK (seasonal limits)	0.10 cm	0.0015 - 0.315	Nedwell and Abram [1978]
Barents Sea	surface	0.53 - 1.44	Ivanov [1978]
Krasnovodsk Bay	surface	0.21 - 0.45	Ivanov [1978]
Coastal			
Limfjord, Denmark	0.10 cm	0.025 - 0.2	Jørgensen [1977*a*]
Black Sea (200 m depth)	surface	0.03 - 0.07	Sorokin [1962]
Aarhus Bay, Denmark	surface	0.44 - 1.01	Jørgensen and Fenchel [1974]
Bay of Kiel	surface	0.008	Hartmann and Nielsen [1969]
St Barbara Basin	0.10 cm	0.005	Kaplam *et al.* [1963]

Station E24, Loch Eil, Scotland	1-5 cm	0.58	C.M. Brown [unpub. data]
E70, Loch Eil, Scotland	1-5 cm	2.80	C.M. Brown [unpub. data]
N.E. Atlantic Stn A3 (150 m depth)	1-5 cm	0.22	C.M. Brown [unpub. data]
Deep Sea			
N.E. Atlantic Stn A2 (2000 m depth)	1.5 cm	0.009	C.M. Brown [unpub. data]
Joides Site 148		2×10^{-7}	Tsou *et al.* [1973]
Cariaco Trench		0.0001	Tsou *et al.* [1973]
Black Sea (2000 m depth)	0.10 cm	0.0008 - 0.04	Sorokin [1962]

and unlikely to be sulphate limited. Generally, the highest rates of sulphate reduction have been reported from sediments of intertidal ecosystems such as salt-marshes, which have high productivity and whose sediments have a large detrital organic content. A progression is noticed with coastal and shelf sediments having intermediate rates of sulphate reduction, while deep oceanic sediments have very low rates. These rates at least partly reflect differences in electron donor availability. Sorokin [1964], for example, has shown that the rates of sulphate reduction are generally much greater in the organically rich slope sediments of the Black Sea than in the profundal sediments whose organic content is low. The high rate of sulphate reduction measured at station E70 in Loch Eil (Table 1) was at a site polluted with highly organic pulp-mill effluent. Ivanov [1978] has calculated that although the total area of the continental shelf is only half that of the continental slope, the total sulphate reduction in shelf sediments is greater because of their higher activity than that in the sediments of the slope (572×10^6 tons S yr^{-1} compared to 375.5×10^6 tons S yr^{-1}).

However, the relationship between overall organic content and sulphate reduction rates is a generalization as sulphate-reducing bacteria are able to directly metabolise only a restricted range of small organic molecules. In natural communities they depend upon the metabolism of other bacterial groups to catalyse the breakdown of detrital material to the molecules which they can use; this is analogous to the similar dependence of methanogenic bacteria [see Wolfe and Higgins, 1979]. As a simple example, Tezuka [1966] has reported syntrophy between fermentative bacteria which metabolized glucose to lactate, which was then utilized by sulphate reducers.

Electron Donors for Sulphate-Reducing Bacteria

Until recently the known sulphate reducing bacteria have been confined to two genera - *Desulfovibrio* and the spore-forming *Desulfotomaculum*. These have been known to metabolize only a restricted range of substrates including lactate, succinate, ethanol, formate and hydrogen [Buchanan and Gibbons, 1974; Postgate, 1979]. Pyruvate is also metabolized in culture, but is unlikely to be excreted into the environment by organotrophic bacteria and is therefore not likely to be a naturally available substrate for sulphate reducers. Recent work has extended our knowledge of possible substrates for sulphate reducers and this is important in helping us to interpret the potential ecological roles of sulphate reducers in natural microbial communities.

Widdel and Pfennig [1977] reported the isolation from both marine and freshwater sediments of *Desulfotomaculum acetoxidans* which oxidized acetate to carbon dioxide, and

Laanbroek and Pfennig [1981] have since isolated a further genus of acetate-oxidizing sulphate reducers which they have called *Desulfobacter*. Such organisms are potentially extremely important as acetate is the major product of anaerobic catabolism of carbon compounds, and the activity of such bacteria will therefore permit the complete anaerobic oxidation of carbon to carbon dioxide. *Desulfobacter* has only been isolated from marine sediments rich in sulphate. Laanbroek and Pfennig [1981] and Widdel [1980] have also reported the isolation of a propionate oxidizing sulphate reducer called *Desulfobulbus*, acetate and carbon dioxide being the end products of propionate catabolism.

Possibly some of the most interesting recent work has been concerned with hydrogen (H_2) metabolism by sulphate reducers. Their ability to utilize H_2 has been appreciated for a considerable time [see Postgate, 1979] but Badziong *et al.* [1978] described a strain of *Desulfovibrio vulgaris* which could use H_2 as its only electron donor, linked to sulphate reduction. Hydrogen metabolism seems to be an important process in regulating the flow of carbon in many anaerobic communities because of interspecies H_2 transfer [see reviews by Wolfe, 1971; Zeikus, 1977; Mah *et al.*, 1977; Wolfe and Higgins, 1979]. The concept of H_2 transfer was established by Bryant *et al.* [1967] who demonstrated that the methanogenic culture known as *Methanobacillus omelianskii* was in fact a syntrophic culture of two bacteria which together converted ethanol to methane:

'S organism' $$CH_3CH_2OH + H_2O \rightarrow CH_3COOH + 2H_2$$

Methanobacterium 'MoH' $$4H_2 + CO_2 \longrightarrow CH_4 + 2H_2O$$

The S organism reoxidized its reduced coenzymes by release of H_2 by the following reaction

$$NADH_2 \rightarrow NAD + H_2$$

This direct release of H_2 is thermodynamically feasible only if the environmental H_2 concentration is very low [Wolin, 1976] and it was the H_2-scavenging activity of the methanogen which maintained this low concentration. Certainly in sediments H_2 seems to be absent or detectable only at very low concentrations [Chen *et al.*, 1972; Winfrey *et al.*, 1977; Strayer and Tiedje, 1978] and turns over extremely rapidly [Winfrey and Zeikus, 1979; Strayer and Tiedje, 1978]. Subsequent work has shown that sulphate reducing bacteria can substitute for methanogenic bacteria as the H_2-scavengers in such associations and indeed where both methanogens and sulphate-reducers are present the sulphate-reducers apparently outcompete the methanogens for the available H_2 if sulphate is also present in the environment [Abram and Nedwell, 1978*a*].

Effect of H_2 Removal in the Microbial Community

Generally, in H_2-transferring associations the H_2-donating partner obtains a slightly greater energy yield if its coenzymes are re-oxidized by H_2-transfer rather than by fermentative reoxidation [Thauer *et al.*, 1977; Tewes and Thauer, 1980]. This is because metabolic intermediates such as pyruvate are not then required to reoxidize reduced coenzymes and can therefore be themselves further oxidized to acetate. A number of workers have demonstrated that in the presence of a H_2-scavenging bacterium a variety of fermentative bacteria show a shift in the balance of their metabolic products from reduced fermentation products such as propionate, succinate, or lactate, with a stoichiometric increase in the amount of acetate and H_2 [for example, Iannotti *et al.*, 1973; Latham and Wolin, 1977]. One may speculate that where there is a limited supply of labile available carbon (that is energy), as in many anaerobic sediments, the small energy gain derived by H_2 transfer may confer a competitive advantage upon H_2 transferring organotrophs over those organotrophs which are obligate fermenters. As a consequence, the production within the community of reduced fermentative products such as lactate, propionate, succinate, etc., may be less important than the production of acetate and H_2. Whether or not this is true will depend upon a variety of environmental parameters, not least being the availability of labile organic carbon. Only further measurements of turnover rates of organic molecules in natural communities will solve this question.

The possible importance of syntrophic interactions such as interspecies H_2 transfer to carbon flow in anaerobic communities has been amplified by the isolation of a variety of bacteria from the rumen and from anaerobic digesters which may also indicate types of organisms present in anaerobic sedimentary communities. *Syntrophobacter wolinii* [Boone and Bryant, 1980] metabolized propionate to acetate, CO_2 and H_2, but only if cocultured with either a methanogen or with a *Desulfovibrio*. Again, anaerobic degradation of benzoate, in the absence of nitrate or light, apparently requires the presence of a consortium which includes a H_2-scavenging organism. Ring cleavage by one microorganism leads to the formation of intermediates including hydrogen. Removal of these intermediates by methanogenic or sulphate-reducing bacteria provides thermodynamically favourable conditions for benzoate degradation [Ferry and Wolfe, 1976; Balba and Evans, 1977; Evans, 1977; Balba and Evans, 1980].

Degradation of even-numbered long chain fatty acids such as butyrate, caproate and caprylate to acetate and H_2; and of odd-numbered long chain fatty acids such as valerate and heptanoate to acetate, propionate, and H_2, was brought about by unnamed bacteria isolated from anaerobic sewage digesters only when cocultured with either a methanogen or sulphate-reducer [McInerney *et al.*, 1979]. Faster growth was obtained when cocultured with the sulphate reducer. In all of these reactions the H_2-scavenging bacteria act as H_2 sinks and permit catabolic reactions by other preceding bacteria which would appear to be thermodynamically impossible in the absence of the H_2 sink.

The flow of carbon in anaerobic sedimentary communities is the result of a complex of microbial interactions analogous to those found in other anaerobic communities such as the rumen and anaerobic digesters. How do sulphate-reducing bacteria appear to function in these sedimentary communities?

Interaction between Sulphate-Reducers and Methanogens

It was reported by a number of workers [Nissenbaum *et al.*, 1972; Claypool and Kaplan, 1974; Martens and Berner, 1977] that methane concentrations in long cores of anoxic marine sediments were generally lowest where sulphate reduction was most active. In a summary of available data Reeburgh and Heggie [1977] showed that in the upper horizons of marine sediments methane concentrations were usually very low, whereas in freshwater sediments methane could be detected throughout the sediment profile, up to the sediment/water interface. They suggested that this could be the result either of competition between methanogens and sulphate-reducers for a common resource, or that sulphate-reducing bacteria might be anaerobically oxidizing methane. In laboratory experiments with jars of marine sediment methanogenesis did not occur until sulphate was depleted [Martens and Berner, 1974]. Cappenberg [1974*a*] found that addition of sulphate to sediment from Lake Vechten, Holland, diminished methane evolution; an observation repeated in sediment from Lake Mendota, USA [Winfrey and Zeikus, 1977]. One proposal to explain this effect [Cappenberg, 1974*b*] was that sulphide derived from sulphate reduction inhibited the methanogenic bacteria. While sulphide can certainly have some effect upon methanogenic bacteria [Cappenberg, 1975; Wellinger and Wuhrmann, 1977; Mountfort and Asher, 1979] evidence to date does not support this hypothesis [Winfrey and Zeikus, 1977; Abram and Nedwell, 1978*b*; Oremland and Taylor, 1978]. Alternatively it was proposed that the two groups of bacteria were competing for either H_2 and/or acetate [Winfrey and Zeikus, 1977]. Addition of H_2 usually stimulated methanogenesis in both marine and freshwater sediments [Winfrey and Zeikus, 1977; Abram and Nedwell, 1978*b*; Oremland and Taylor, 1978] but sulphate reduction was also simultaneously stimulated [Abram and Nedwell, 1978*b*]. In a laboratory culture simulation, sulphate-reducing bacteria from saltmarsh sediment outcompeted a

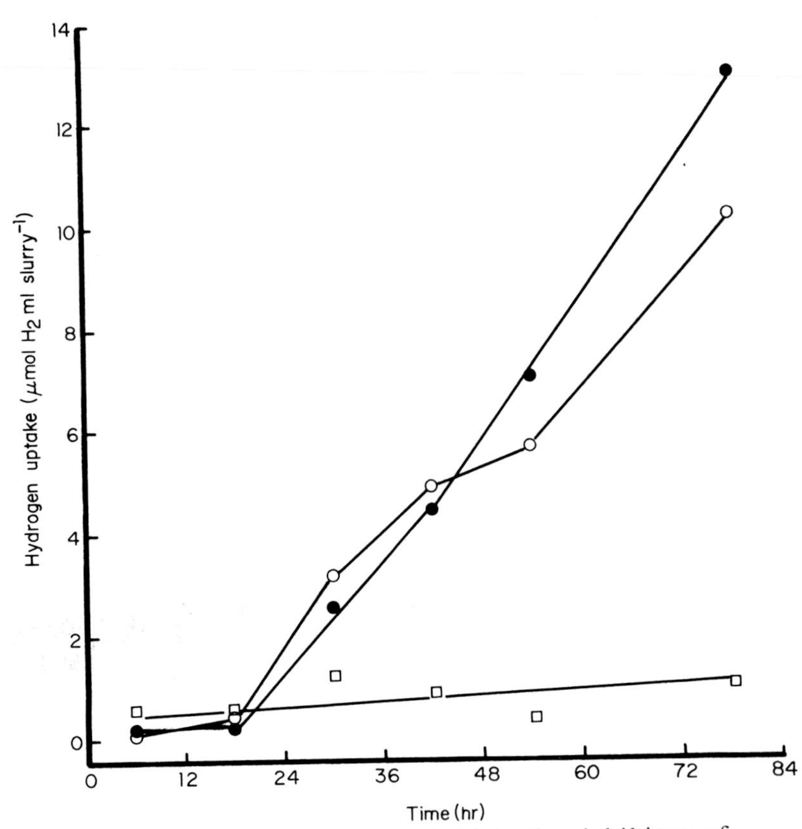

Fig. 3 The effect of addition of molybdate (an inhibitor of sulphate-reducing bacteria) and chloroform (an inhibitor of methanogenic bacteria) upon removal of H_2 from the gas head space above a slurry of saltmarsh sediment incubated under a H_2 (80% v/v): CO_2 (20% v/v) atmosphere. ● = control, ○ = + chloroform (0.01% v/v), □ = + chloroform + molybdate (20 mmol l^{-1}) [Banat and Nedwell, unpub. data].

methanogen for transferred hydrogen [Abram and Nedwell, 1978*a*]. In experiments with slurries of sediment from both saltmarsh (Fig. 3) and tropical sea grass beds [Oremland and Taylor, 1978] the presence of 20 mM molybdate, an inhibitor of sulphate-reducing bacteria [Peck, 1959; Oremland and Taylor, 1978; Andreesen, 1980] largely inhibited uptake of hydrogen from the headspace above the slurry while the presence of chloroform, an inhibitor of methanogenic bacteria [Bauchop, 1967] had little effect. Mountfort *et al.* [1980] also found that H_2 uptake by slices of sediment from Delaware Inlet, New Zealand, appeared to be linked to sulphate reduction and not methanogenesis. So it would seem that in marine sediments the sulphate-reducing bacteria, not the methanogens, are the major H_2-scavengers.

It is interesting that in a further slurry experiment [Banat and Nedwell, unpub. data] the addition of 40 mM lactate to slurry also inhibited H_2 uptake from the head-space during the period that lactate persisted, but the H_2 uptake resumed once lactate was diminished (Fig. 4). This suggested that reduced fermentative intermediates

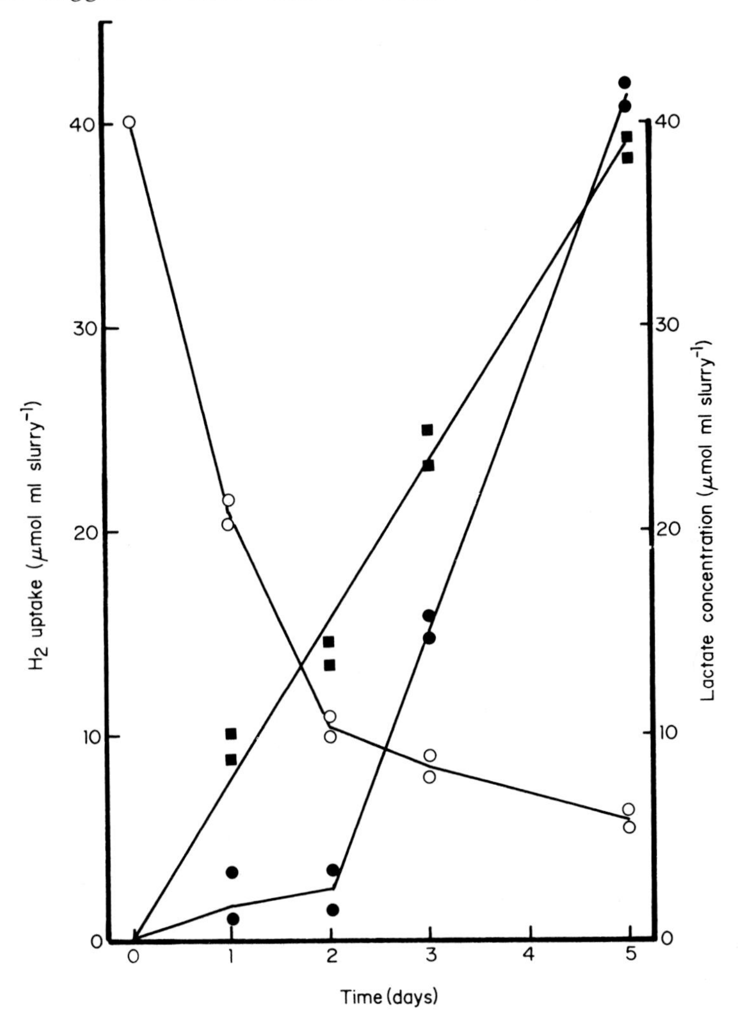

Fig. 4 The effect of H_2 uptake from the gas headspace of the addition of lactate (40 mmol l^{-1}) to a slurry of saltmarsh sediment incubated under a H_2 (80% v/v): CO_2 (20% v/v) atmosphere. ■ = H_2 uptake in control, ● = H_2 uptake in treatment with added lactate (40 mmol l^{-1}), ○ = residual lactate in treatment with added lactate [Banat and Nedwell, unpub. data].

such as lactate were used preferentially to H_2 by the sulphate-reducing bacteria if they were present in the environment. As discussed previously, such reduced fermentative intermediates are less likely to be produced if H_2 transfer occurs, and the importance of sulphate-reducers as H_2-scavengers would, therefore, appear to reflect an environment which is carbon/energy limited. At first sight this may seem peculiar as many of these intertidal sedimentary environments may have high organic contents, but in the saltmarsh this organic material may be refractory and only slowly broken down [Abd Aziz and Nedwell, 1979] so that the supply of labile available carbon remains restricted. Lactate has been reported in the marine environment, for example in an organically rich cellulose-polluted Scottish loch sediment [Miller *et al.*, 1979], and also in sediment in Lake Vechten, Holland [Cappenberg, 1974*b*]. In the latter case the presence of lactate may reflect the input of relatively labile carbon from planktonic primary producers, rather than the more refractile detritus of higher plants. In sediment from the salt marsh at Colne Point, Essex, UK the predominant fatty acids are acetate, propionate and butyrate, and lactate is not detectable.

Acetate is a major end-product of anaerobic carbon oxidation [see Wolfe, 1971] and may be a possible precursor of methane. To take some examples, Cappenberg and Prins [1974] calculated that 70% of the methane in Lake Vechten was from acetate; a maximum of 59% from acetate in Lake Mendota [Winfrey and Zeikus, 1979]; 33% in Blelham Tarn, UK [Jones, this volume]; and a maximum varying between 2 to 69% in a number of Russian lakes [Belyaev *et al.*, 1975]. The balance of methane has generally been assumed to come from the $H_2 + CO_2$ pathway with which H_2-scavenging sulphate-reducers compete. However, in addition to competition between methanogens and sulphate-reducers for H_2, the presence of sulphate also appears to have a profound effect upon the fate of acetate in anaerobic sediments.

Acetate Metabolism in Anaerobic Sediments

In Lake Mendota sediments, acetate metabolism varied seasonally and appeared to be related to the concentration of sulphate in the water [Winfrey and Zeikus, 1979]. If [$2^{14}C$]acetate was added to sediment samples, on average 42% of the label was recovered as $^{14}CH_4$ and 20% as $^{14}CO_2$, with the balance being assimilated within the sediment. In contrast, in the bottom of the water column where sulphate was present the methyl carbon of acetate was oxidized to $^{14}CO_2$. Corroborating results have been reported from other environments. Mountfort *et al.* [1980] found that in a New Zealand marine coastal sediment, [$2^{14}C$]acetate was entirely oxidized to $^{14}CO_2$, except for one site where sulphate concentrations within the sediment were low, and here $^{14}CH_4$ was formed. Again, in a US *Spartina* saltmarsh sediment acetate was oxidized to carbon dioxide at a site where sulphate concentrations were always high, but methane became more important as a product at a site where sulphate concentrations were lower

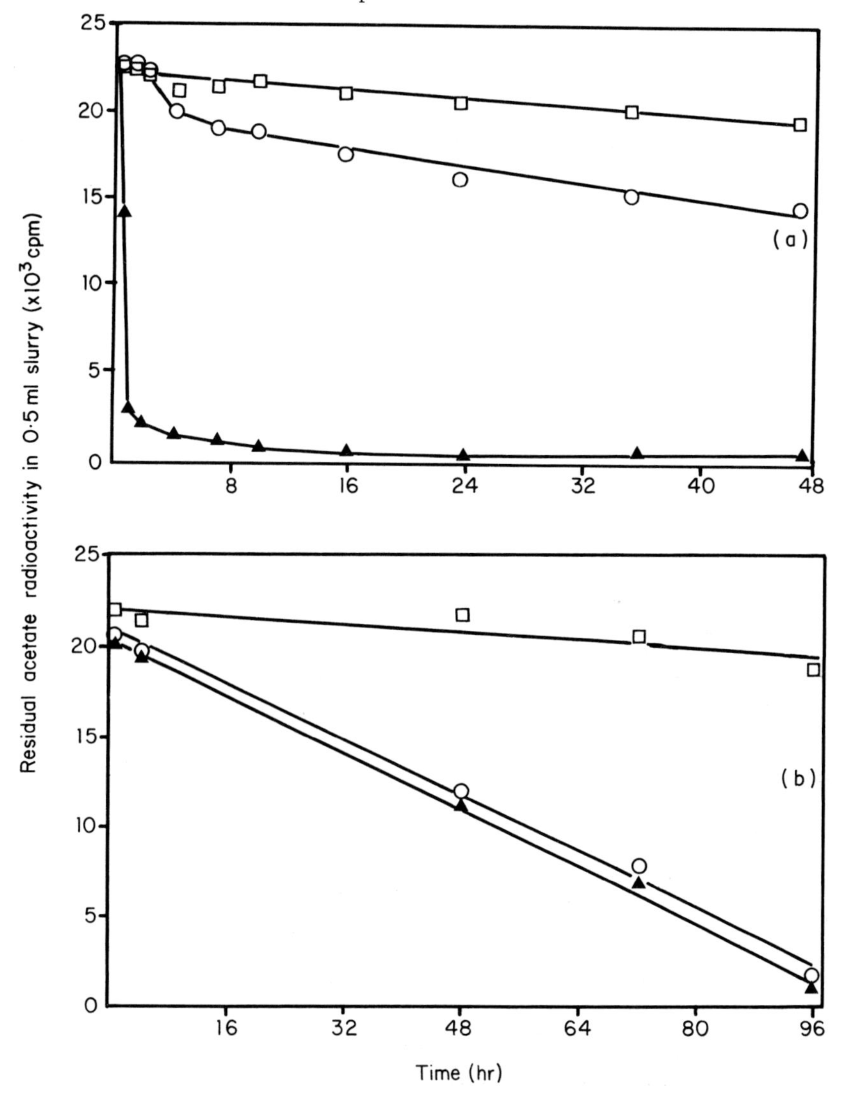

Fig. 5 Effect of addition of molybdate (20 mmol l^{-1}) and fluoroacetate (5 mmol l^{-1}) upon the turnover of [U^{14}C]acetate in slurries of sediment under N_2 (80% v/v): CO_2 (20% v/v).
a) saltmarsh sediment **b)** freshwater lake sediment ▲ = residual acetate radioactivity in control, O = residual acetate radioactivity in slurry + molybdate, □ = residual acetate radioactivity in slurry + fluoroacetate [from Banat, Lindström, Nedwell and Balba, in prep.].

[King and Wiebe, 1980]. Similar results have been found in the Colne Point saltmarsh. The complete oxidation of acetate to carbon dioxide would, therefore, appear to suggest the presence of acetate-oxidizing sulphate-reducers. Some experiments with slurry of saltmarsh sediment tended to support this suggestion. Molybdate was found to inhibit the oxidation of uniformly labelled ^{14}C acetate to $^{14}CO_2$ in saltmarsh sediment; but this inhibitor of sulphate-reducing bacteria had no effect upon acetate turnover in freshwater sediment devoid of sulphate (Fig. 5). In both cases, though, acetate metabolism was

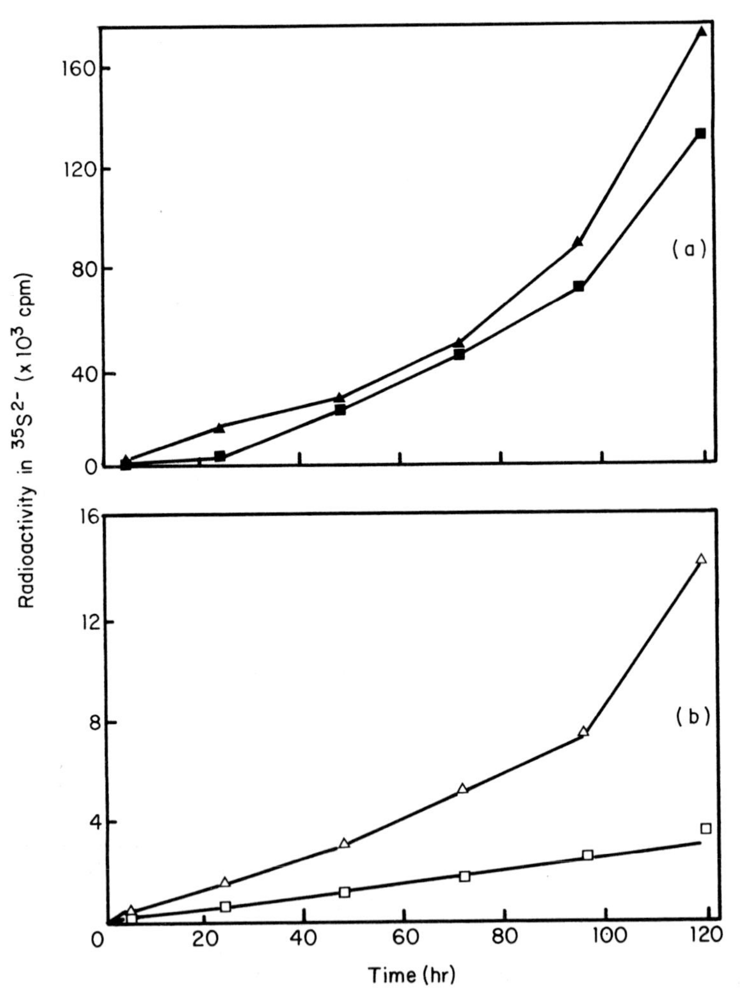

Fig. 6 Effect of fluoroacetate upon the rate of sulphate reduction in saltmarsh sediment slurry. **a)** slurry under H_2 (80% v/v): CO_2 (20% v/v) **b)** slurry under N_2 (80% v/v): CO_2 (20% v/v). Triangles show controls with no additions: squares are slurry + fluoroacetate (5 mmol l^{-1}).

inhibited by the addition of fluoroacetate (5 mM), an inhibitory analogue of acetate [Cappenberg, 1974*b*; Winfrey and Zeikus, 1979].

In a following experiment it was found that the addition of fluoroacetate to saltmarsh sediment slurry partially inhibited the rate of ^{35}S sulphate reduction. The proportionate inhibition was less under a $H_2:CO_2$ atmosphere than under a $N_2:CO_2$ atmosphere (Fig. 6). This was interpreted as showing the presence of two distinct functional groups of sulphate-reducing bacteria, one population metabolizing H_2 and not being inhibited by fluoroacetate. The second population is of acetate-ozidizing sulphate-reducers which are inhibited by fluoroacetate, and which do not metabolize H_2. Under a H_2 atmosphere the former group is stimulated and the proportionate inhibition of $^{35}SO_4$ reduction by fluoroacetate is therefore diminished. Both of the described genera of acetate-oxidizing sulphate-reducers are unable to metabolize H_2 [Widdel and Pfennig, 1977; Laanbroek and Pfennig, 1981].

Thus, the sulphate-reducing bacteria in the Colne Point saltmarsh sediment seem to consist of at least two functional groups, each successfully competing for one of the major end products of anaerobic carbon oxidation - that is, either H_2 or acetate. Each of these molecules is, of

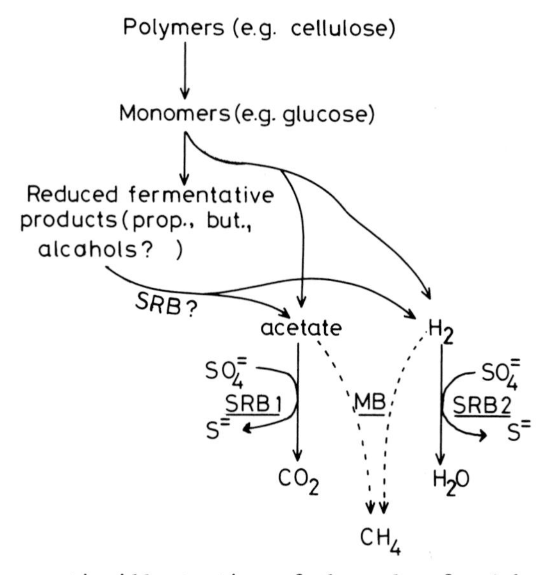

Fig. 7 Diagrammatic illustration of the role of sulphate-reducing bacteria (SRB) in carbon/electron flow in anaerobic marine sediment. The sites of potential competition for common substrates with methanogenic bacteria (MB) are indicated by broken lines, showing that MB are usually outcompeted by SRB when sulphate is present. SRB1 = acetate-oxidizing SRB, SRB2 = H_2-oxidizing SRB.

course, also a potential precursor of methane, and sulphate-reducing bacteria and methanogenic bacteria therefore seem to occupy similar biochemical niches. The sulphate reducers outcompete the methanogens, unless their activity is limited by low sulphate availability. As a consequence, while methane may be the major end-product of anaerobic carbon mineralization in freshwater sediments, the anaerobic oxidation of organic carbon to carbon dioxide via acetate by sulphate-reducing bacteria seems to be the major pathway in marine and intertidal sediments (Fig. 7). A transition between these two situations has been described in the White Oak River estuary, N. Carolina [Martens and Goldhaber, 1978] where in riverbed stretches subjected to seawater influence sulphate reduction in the sediment, as shown by the presence of pyrite, predominated. Upstream, away from the marine influence, methane generation predominated.

Both sulphate-reducers and methanogens depend upon preceding bacteria to provide their energy substrates; but equally the removal of these end-products by either sulphate-reducer or methanogen may have an important ecological function in thermodynamically "pulling" the anaerobic catabolism of otherwise relatively refractile molecules, in the manner already demonstrated in laboratory cultures. It may be that other functional groups of sulphate-reducers capable of directly oxidizing other fatty acids and alcohols also exist in the sediment so that there is a direct metabolism of these intermediates by sulphate-reducers in addition to a kinetic effect because of the scavenging of H_2 and acetate. The propionate-oxidizing sulphate-reducers reported by Laanbroek and Pfennig [1981] are an example.

Methanogenesis may not be entirely inhibited by sulphate reduction, but a competitive balance exists between the two processes and methanogenesis proceeds very slowly in comparison with sulphate reduction. Oremland and Taylor [1978] have shown that methanogenesis proceeds slowly even in the presence of active sulphate reduction, and methanogenesis from $H^{14}CO_3$ occurred in the same surface strata of UK saltmarsh sediment where sulphate reduction was detected [Senior *et al.*, in prep.]. Figure 8 illustrates the profiles of the mean annual rates of sulphate reduction and bicarbonate methanogenesis in the top 20 cm of sediment profile, and the two processes concurred although methanogenesis was about three orders of magnitude slower than sulphate reduction. Methanogenesis also occurred during the same summer period when sulphate reduction was most active (Fig. 9), and there was therefore no apparent vertical or temporal separation of the two processes over these shallow cores of sediment. Calculations of microenvironment size based upon the example given by Jørgensen [1977*a*] have tended to suggest that this apparent coexistence is not the result of sulphate-depleted microenvironments in

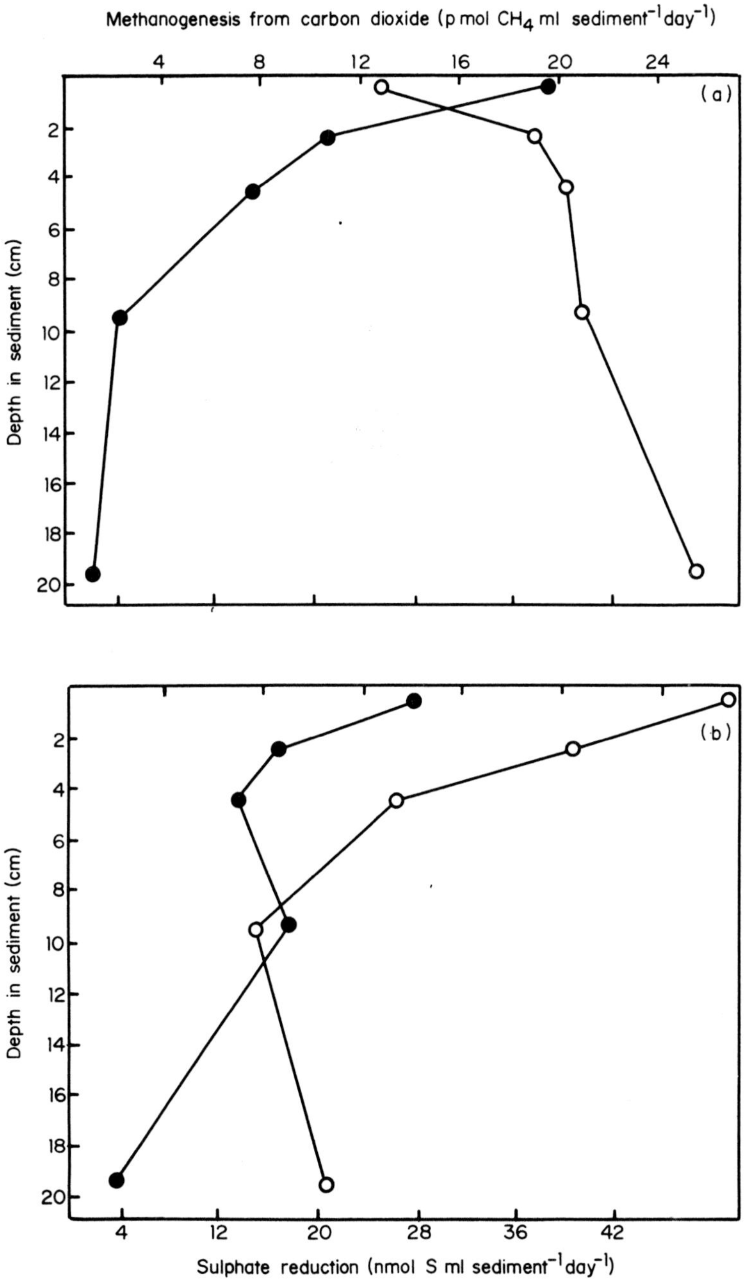

Fig. 8 Profiles of the annual mean rates of sulphate reduction (●) and methanogenesis from carbon dioxide (○) with depth in sediments from **a)** saltmarsh pan **b)** saltmarsh creek. [From Senior, Lindström, Banat and Nedwell, in prep.].

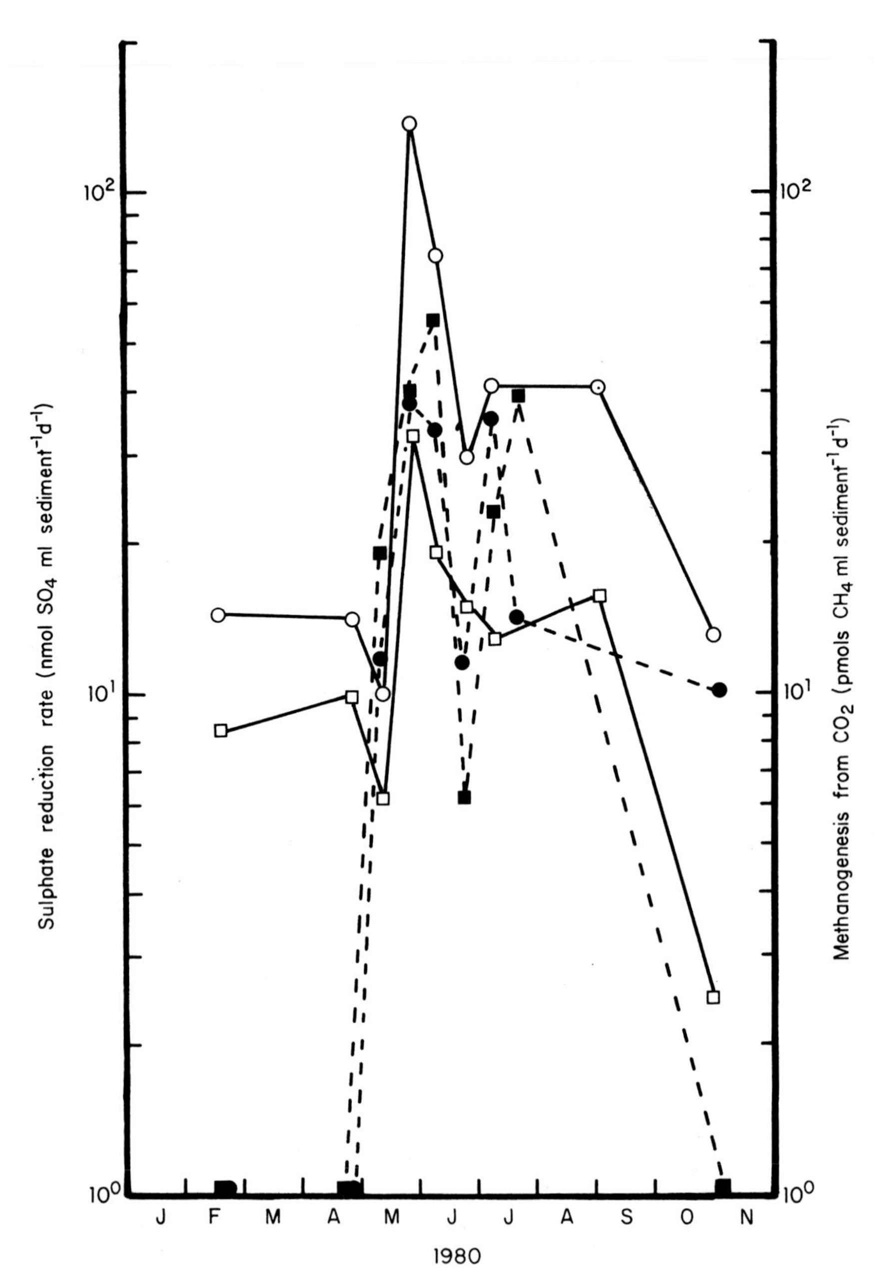

Fig. 9 Seasonal changes in the rates of sulphate reduction (open symbols) and methanogenesis from carbon dioxide (closed symbols) in sediment from a saltmarsh pan. Circles = 0 to 1 cm depth; squares = 9 to 10 cm depth.

which methanogenesis can proceed in isolation from competition with sulphate reducers [Senior *et al.*, in prep.]. Rather, it seems that methanogens persist within the same environment as the sulphate-reducers even though obtaining only a very small proportion of the common environmental resource. Unlike a chemostat, outcompeted bacteria will not be washed out from a sediment and may therefore persist, although at extremely low rates of activity compared to their more successful competitor. In deeper layers of sediment where sulphate-reducers become limited, for example by depleted sulphate, the thermodynamic balance changes and methanogenesis predominates over sulphate reduction.

The depleted concentrations of methane in the surface layers of marine sediments where sulphate reduction is active may also be partly the result of anaerobic reoxidation of methane by sulphate-reducers, as suggested by Reeburgh and Heggie [1977]. Reeburgh [1980] has recently confirmed, using $^{14}CH_4$, that methane is oxidized to $^{14}CO_2$ throughout the sulphate reduction zone in Skan Bay sediment, reaching a maximum rate near the base of the sulphate-reduction zone.

Sulphide from Organic Matter

Sulphide may originate by the dissimilatory reduction of sulphate, but can also be formed putrefactively by the anaerobic chemotrophic metabolism of thio-organic compounds such as cysteine or methionine [for example, Zinder and Brock, 1978]. The importance of each source to total sulphide formation varies in different sediments. Organically derived sulphide is usually relatively unimportant in marine sediments [Nedwell and Floodgate, 1972*a*; Jørgensen and Fenchel, 1974], although its importance will increase if sulphate becomes limiting for dissimilatory reduction [Gunkel and Oppenheimer, 1963; Nedwell and Floodgate, 1972*a*]. In contrast, in freshwater sediments where sulphate is normally present in comparatively low concentrations a major proportion of the sulphide may be derived putrefactively from organically-bound sulphur. Nriagu [1968] estimated that 45% of sulphide in the sediment of Lake Mendota was from this source. Molongoski and Klug [1980] also suggested that putrefactive formation of sulphide was more important than sulphate dissimilation in Wintergreen lake, USA; and Jones [1980] estimated roughly 80% of the sulphide in Blelham Tarn was from organic sulphur.

Sinks for Sulphide

Precipitation of Sulphide Minerals

Sulphide formed within the sediment can meet a variety of fates. Firstly, many metal sulphides are relatively insoluble and may be precipitated as solid sulphide minerals within the sediment. The sulphides of iron are the most important as there is a high iron content in many sediments [Cronan, 1974; Goldhaber and Kaplan, 1974]. Usually the amorphous, poorly crystalline mineral hydrotroilite ($FeS.nH_2O$) is the first product [Berner, 1964]. This may be diagenetically converted to the more stable mineral pyrite (FeS_2) by addition of sulphur [Berner, 1970].

$$Fe^{2+} + HS^- \rightarrow FeS + H^+$$

$$FeS + S^o \rightarrow FeS_2$$

In freshwater sediment at high pH, marcasite rather than pyrite may be formed. Pyrite can be formed directly without the intervention of hydrotroilite [Berner, 1964; Goldhaber and Kaplan, 1974] under conditions of low pH (5 to 6.5) and where iron monosulphides are undersaturated. Howarth and Teal [1979] have demonstrated that pyrite is rapidly formed in this manner in the peat of the Great Sippewissett marsh, Cape Cod, USA, and there were seasonal variations in the sedimentary pyrite content. Pyrite was oxidized around *Spartina* roots during summer, but accumulated during winter when the grass was inactive. However, gradual diagenetic conversion of hydrotroilite to pyrite seems to be the most common situation, reflected by trends of decrease of hydrotroilite with depth (that is, with time) but increase of pyrite as it is formed diagenetically from hydrotroilite [for example, Kaplan *et al.*, 1963; Berner, 1970].

Loss of Dissolved Sulphide from Sediment

Sulphide which is not precipitated will remain dissolved in the pore water, and the concentration of dissolved sulphide will therefore be greatly influenced by the availability of precipitable iron [Nedwell and Abram, 1978]. The dissolved sulphide may be transported within the sediment in response to diffusion gradients and to physical transport such as bioturbation or tidal flushing. In many coastal and intertidal sediments the aerobic zone is confined to only the surface few millimetres depth [Revsbech *et al.*, 1980] and the zone of most active sulphate reduction is very near to the sediment/water interface. Consequently, dissolved sulphide may rapidly reach the surface and be lost from the sediment. Thus the biologically active surface layers of sediment tend to be very "open" with sulphate diffusing in and sulphide diffusing out, so that the total turnover of sulphate vastly exceeds that of just the sulphate in the seawater initially entrained when the sediment was deposited. For example, Jørgensen [1977*c*] estimated for a Danish coastal sediment that > 90% of the sulphide formed annually within the top

20 cm of sediment was lost again to the seawater, and a similar figure was calculated for sulphide loss from a saltmarsh sediment [Nedwell and Abram, 1978].

Thus, while sulphate reduction accounts for a considerable proportion of the organic carbon mineralization in marine sediments, the product, sulphide, tends to be rapidly lost from the sediment, at least from the biologically active surface layers. This is important as sulphide is thermodynamically unstable in aerobic environments and represents a potential source of energy. Howarth and Teal [1980] estimated that 29% of the primary production of the Great Sippiwissett saltmarsh was exported from the ecosystem, 22% of the energy being in the form of reduced sulphur compounds. It has also been suggested that sulphide emitted to the atmosphere from anaerobic environments accounts for the apparent imbalance in the global atmospheric sulphur cycle, where calculated removal of sulphur exceeds apparent additions [Kellog *et al.*, 1972].

Reoxidation of Sulphide

Chemical reoxidation In the presence of oxygen, sulphide may be chemically reoxidized [Chen and Morris, 1972; Almgren and Hagström, 1974]. Studies of sulphide reoxidation in seawater suggested that the initial oxidation was chemical, forming thiosulphate [Sorokin, 1972; Almgren and Hagström, 1974; Goldhaber and Kaplan, 1974].

$$H_2S + O_2 \rightarrow 2S^o + 2H_2O$$

$$H_2S + 2O_2 \rightarrow SO_3^{2-} + H_2O$$

$$SO_3^{2-} + S^o \rightarrow S_2O_3^{2-}$$

Biological oxidation was subsequently responsible for the further oxidation of thiosulphate to sulphate [Sorokin, 1972; Tuttle and Jannasch, 1973].

Thus, while sulphide may be reoxidized when it diffuses into aerobic regions the mechanism(s) of its reoxidation can vary. Beside the possibility of either chemical or biological reoxidation, the pathways of biological oxidation can vary markedly depending upon the microorganisms catalysing the reoxidation. The microorganisms present in turn depend largely upon the particular environmental conditions.

Biological reoxidation Chemolithotrophic bacteria may utilize sulphide, thiosulphate or other inorganic reduced sulphur compounds, as electron donors [Aleem, 1975; Kuenen, 1975]. All of these bacteria are obligate aerobes, with the exception of *Thiobacillus denitrificans* which can utilize nitrate. In general, therefore, they occur at the interface between aerobic layers of sediment from which oxygen is transported, and anaerobic layers of sediment from which sulphide diffuses. Where such opposing diffusion gradients overlap tends to be the site of high bacterial activity. For example, Sorokin [1972] showed that in the water column of the Black Sea a peak of $^{14}CO_2$-fixing microbial activity occurred in the redox discontinuity zone where gradients of sulphide and oxygen overlapped, and this lithotrophic activity was attributed to the activity of thiobacilli. Again, Kepkay and Novitsky [1980] have reported maximum carbon dioxide fixation at the aerobic/anaerobic interface at 40 cm depth in sediment from Halifax Harbour, Nova Scotia. A peak of sulphate concentration at this same depth indicated that sulphide-oxidizing bacteria were at least partially responsible for this activity. Indeed, the organic carbon resulting from such fixation may represent an important input into the sedimentary system [Kepkay *et al.*, 1979].

The thiobacilli are not the only sulphide-oxidizing lithotrophic bacteria. Recent investigations have revealed the widespread distribution of the large filamentous sulphide-oxidizing bacteria of the genus *Beggiatoa*, and related genera. They have been reported from both marine and freshwater sediments [Jørgensen, 1977*b*; Gallardo, 1977; Stanley *et al.*, 1978; Jones, this volume], and may be important in the reoxidation of sulphide in marine sediments where their population density can be high [Jørgensen, 1977*b*]. However, there have been no quantitative estimates of the relative importance of these two groups of chemolithotrophs to sulphide oxidation.

Photosynthetic reoxidation of sulphide is also possible in those sediments which lie within the euphotic zone. Such reoxidation is the result of the activity of two groups of photosynthetic bacteria [Pfennig, 1975] which use hydrogen sulphide as an electron donor for photosynthesis. Also recent work has revealed that at least some blue-green algae (cyanobacteria) are capable under anaerobic conditions of anoxygenic photosynthesis where hydrogen sulphide is used as an electron donor and sulphur granules are deposited, as in bacterial photosynthesis [Cohen *et al.*, 1975; Garlick *et al.*, 1977]. Pfennig [1975] has suggested that in environments where both photolithotrophic and chemolithotrophic bacteria are active the former should predominate as they are able to synthesize four times more cell material per mole of sulphide than chemolithotrophs such as *Beggiatoa*. Below the euphotic zone, though, only chemolithotrophic reoxidation of sulphide is possible.

The sulphide-oxidizing phototrophs are also gradient dwellers, occurring where sulphide is present in illuminated environments. This essentially confines photosynthetic sulphide reoxidation to the thin surface layer of sediment (a few mm only) into which light penetrates

[Fenchel and Straarup, 1971]. Blackburn *et al.* [1975] illuminated the surface of sediment cores in the laboratory and electrochemically measured sulphide distributions within the sediment. During periods of illumination the dissolved sulphide concentrations in the immediately subsurface layers of sediment decreased markedly as sulphide was reoxidized photosynthetically, and then the concentrations of sulphide recovered during dark periods. The benthic flora of photosynthetic microorganisms was therefore an obviously effective sink for sulphide, and a possible barrier for diffusion of sulphide into the overlying water. This was supported by the field study of Hansen *et al.* [1978] where two coastal sediments covered only by a few cm of water were examined for release of hydrogen sulphide. Emission of sulphide into the water occurred only at night when the benthic mat of photosynthetic bacteria and blue-green algae was light-limited. During the day the benthic mat effectively prevented loss of sulphide from the surface of the sediment. Even at night when sulphide was released, chemical oxidation of sulphide occurred within the water column above the sediment. Östlund and Alexander [1963] calculated that hydrogen sulphide was likely to be chemically oxidized before reaching the atmosphere if released into a water column of more than a few metres depth. However, Hansen *et al.* [1978] concluded that their shallow water coastal sediments could, over 24 hours, be net exporters of hydrogen sulphide to the atmosphere. In the marine sediments so far examined it must be concluded that although there may be almost complete loss of annually formed sulphide from the sediment it is not likely to be lost as sulphide but rather reoxidized in the uppermost layers of sediment and in the immediately adjacent water by either chemical or biological means. The loss of sulphur from freshwater sediments may be considerably lower. Nriagu and Coker [1976] estimated that only about 1% of the annual sulphur input into Lake Ontario sediment was subsequently reemitted from the sediment.

Although it has been concluded [Kellog *et al.*, 1972] that emission of sulphide from most marine sediments is unlikely to be the "missing source" of atmospheric sulphur, it is nevertheless possible that intertidal sediments are exporters of hydrogen sulphide to the atmosphere. Volatile organic sulphide compounds have also been suggested as a possible source of atmospheric sulphur. Lovelock *et al.* [1972] detected measurable levels of dimethyl sulphide in Atlantic water but Liss and Slater [1974] calculated that the flux across the seawater/atmosphere interface was only about 4% of the influx of sulphur required to balance the atmosphere sulphur budget. However, it is known that a variety of other organic-sulphide compounds are also formed in soils [Francis *et al.*, 1975] and aquatic sediments [Zinder and Brock, 1978], including dimethyl sulphide, dimethyl disulphide, methane thiol, and *n*-propane thiol. These are formed by the bacterial metabolism of sulphur-containing organic compounds, and are therefore unlikely to be quantitatively significant in the marine environment compared to sulphide, but may be important in freshwater sediments. There is no information on their distribution in aquatic sediments. Whether there is a significant input of such volatile organic sulphide compounds from sediments to the atmosphere is unknown and requires further investigation.

References

Abd. Aziz, S.A. and Nedwell, D.B. (1979). Microbial nitrogen transformations in the salt marsh environment. In "Ecological Processes in Coastal Environments" (Eds. R.L. Jefferies and A.J. Davy), pp.385-398. Blackwell, Oxford.

Abdollahi, H. and Nedwell, D.B. (1979). Seasonal temperature as a factor influencing bacterial sulphate reduction in a saltmarsh sediment. *Microbial Ecology* **5**, 73-79.

Abram, J.W. and Nedwell, D.B. (1978*a*). Inhibition of methanogenesis by sulphate reducing bacteria competing for transferred hydrogen. *Archives of Microbiology* **117**, 89-92.

Abram, J.W. and Nedwell, D.B. (1978*b*). Hydrogen as a substrate for methanogenesis and sulphate reduction in anaerobic saltmarsh sediment. *Archives of Microbiology* **117**, 93-97.

Aleem, M.I.H. (1975). Biochemical reaction mechanisms in sulphur oxidation by chemosynthetic bacteria. *Plant and Soil* **43**, 587-607.

Almgren, T. and Hagström, I. (1974). The oxidation rate of sulphide in seawater. *Water Research* **8**, 395-400.

Andreesen, J.R. (1980). Role of selenium, molybdenum and tungsten in anaerobes. In "Anaerobes and Anaerobic Infections" (Eds. G. Gottschalk, N. Pfennig and H. Werner), pp.31-39. Gustav Fischer Verlag, Stuttgart.

Badziong, W., Thauer, R.K. and Zeikus, J.G. (1978). Isolation and characterization of *Desulfovibrio* growing on hydrogen plus sulphate as the sole energy sources. *Archives of Microbiology* **116**, 41-49.

Balba, M.T. and Evans, W.C. (1977). The methanogenic fermentation of aromatic substrates. *Biochemical Society Transactions* **5**, 302-304.

Balba, M.T. and Evans, W.C. (1980). The anaerobic dissimilation of benzoate by *Pseudomonas aeruginosa* coupled with *Desulfovibrio vulgaris*, with sulphate as terminal electron acceptor. *Biochemical Society Transactions* **8**, 624-625.

Bauchop, T. (1967). Inhibition of rumen methanogenesis by methane analogues. *Journal of Bacteriology* **94**, 171-175.

Belyaev, S.S., Finkel'shtein, Z.I. and Ivanov, M.V. (1975). Intensity of bacterial methane formation in ooze deposits of certain lakes. *Microbiology* **44**, 272-275.

Berner, R.A. (1964). Iron sulphides formed from aqueous solutions at low temperatures and atmospheric pressure. *Journal of Geology* **72**, 293-306.

Berner, R.A. (1970). Sedimentary pyrite formation. *American Journal of Science* **268**, 1-23.

Berner, R.A. (1974). Kinetic models for the early diagenesis of nitrogen, sulphur, phosphorus and silicon in anoxic marine sediments. In "The Sea" (Ed. E.D. Goldberg) Vol.5, pp.427-450. Wiley-Interscience, New York.

Blackburn, T.H., Kleiber, P. and Fenchel, T. (1975). Photosynthetic sulphide oxidation in marine sediments. *Oikos* **26**, 103-108.

Boone, D.R. and Bryant, M.P. (1980). Propionate-degrading bacterium, *Syntrophobacter wolinii* sp. nov., gen. nov., from methanogenic ecosystems. *Applied and Environmental Microbiology* **40**, 626-632.

Bryant, M.P., Wolin, E.A., Wolin, M.J. and Wolfe, R.S. (1967). *Methanobacillus omelianskii*, a symbiotic association of two species of bacteria. *Archives für Mikrobiologie* **59**, 20-31.

Buchanan, R.E. and Gibbons, N.E. (Eds.) (1974). Bergey's Manual of Determinative Bacteriology, Eighth edition. Williams and Wilkins, Baltimore.

Cappenberg, T.E. (1974*a*). Interrelations between sulphate-reducing and methane-producing bacteria in bottom deposits of a freshwater lake. I. Field observations. *Antonie van Leeuwenhoek* **40**, 285-295.

Cappenberg, T.E. (1974*b*). Interrelations between sulphate-reducing and methane-producing bacteria in bottom deposits of a freshwater lake. II. Inhibition experiments. *Antonie van Leeuwenhoek* **40**, 297-306.

Cappenberg, T.E. (1975). A study of mixed continuous cultures of sulphate-reducing and methane-producing bacteria. *Microbial Ecology* **2**, 60-72.

Cappenberg, T.E. and Prins, R.A. (1974). Interrelations between sulphate-reducing and methane-producing bacteria in bottom deposits of a freshwater lake. III. Experiments with ^{14}C-labelled substrates. *Antonie van Leeuwenhoek* **40**, 457-469.

Chen, K.Y. and Morris, J.C. (1972). Kinetics of oxidation of aqueous sulphide by O_2. *Environmental Science and Technology* **6**, 529-537.

Chen, R.L., Keeney, D.R., Konrad, J.G., Holding, A.J. and Graetz, D.A. (1972). Gas metabolism in sediments of Lake Mendota. *Journal of Environmental Quality* **1**, 155-158.

Claypool, G.E. and Kaplan, I.R. (1974). The origin and distribution of methane in marine sediments. In "Natural Gases in Marine Sediments" (Ed. I.R. Kaplan), pp.99-139. Plenum Press, New York.

Cohen, Y., Padan, E. and Shilo, M. (1975). Facultative anoxygenic photosynthesis in the cyanobacterium *Oscillatoria limnetica*. *Journal of Bacteriology* **123**, 855-861.

Cronan, D.S. (1974). Authigenic minerals in deep sea sediments. In "The Sea" (Ed. E.D. Goldberg), Vol.5, pp.491-525. Wiley-Interscience, New York.

Deuser, W.G. (1975). Reducing environments. In "Chemical Oceanography" (Eds. J.P. Riley and G. Skirrow), pp.1-37. Academic Press, London.

Evans, W.C. (1977). Biochemistry of the bacterial catabolism of aromatic compounds in anaerobic environments. *Nature* **270**, 17-22.

Fenchel, T. and Straarup, B.J. (1971). Vertical distribution of photosynthetic pigments and the penetration of light in marine sediments. *Oikos* **22**, 172-182.

Ferry, J.G. and Wolfe, R.S. (1976). Anaerobic degradation of benzoate to methane by a microbial consortium. *Archives of Microbiology* **107**, 33-40.

Francis, A.J., Duxbury, J.M. and Alexander, M. (1975). Formation of volatile organic products in soils under anaerobiosis. II. Metabolism of amino acids. *Soil Biology and Biochemistry* **7**, 51-56.

Gallardo, V.A. (1977). Large benthic microbial communities in sulphide biota under Peru-Chile Subsurface Countercurrent. *Nature* **268**, 331-332.

Garlick, S., Oren, A. and Padan, E. (1977). Occurrence of facultative anoxygenic photosynthesis among filamentous and unicellular cyanobacteria. *Journal of Bacteriology* **129**, 623-629.

Goldhaber, M.B. and Kaplan, I.R. (1974). The sulphur cycle. In "The Sea" (Ed. E.D. Goldberg), pp.569-655. Wiley, New York.

Gunkel, W. and Oppenheimer, C.H. (1963). Experiments regarding the sulphide formation in sediments of the Texas Gulf Coast. In "Symposium on Marine Microbiology" pp.674-683. Charles C. Thomas, Springfield.

Hansen, M.H., Ingvorsen, K. and Jørgensen, B.B. (1978). Mechanisms of hydrogen sulphide release from coastal marine sediments to the atmosphere. *Limnology and Oceanography* **23**, 68-76.

Hartmann, M. and Nielsen, H. (1969). δ^{34}S-Werte in rezenten Meeressedimenten und ihre Deutung am Beispiel einiger Sedimentprofile aus der westlichen Ostsee. *Geologische Rundschau* **58**, 621-655.

Howarth, R.W. and Teal, J.M. (1979). Sulphate reduction in a New England saltmarsh. *Limnology and Oceanography* **24**, 999-1013.

Howarth, R.W. and Teal, J.M. (1980). Energy flow in a salt marsh ecosystem: the role of reduced inorganic sulphur compounds. *American Naturalist* **116**, 862-872.

Iannotti, E.L., Kafkewitz, C., Wolin, M.J. and Bryant, M.P. (1973). Glucose fermentation products of *Ruminococcus albus* grown in continuous culture with *Vibrio succinogenes:* changes caused by interspecies transfer of H_2. *Journal of Bacteriology* **114**, 1231-1240.

Ivanov, M.V. (1978). Influence of microorganisms and microenvironment on the global sulphur cycle. In "Environmental Biogeochemistry and Geomicrobiology" (Ed. W.E. Krumbein), pp.47-61. Ann Arbor Science Publishers, Michigan.

Jones, J.G. (1976). The microbiology and decomposition of seston in open water and experimental enclosures in a productive lake. *Journal of Ecology* **64**, 241-278.

Jones, J.G. and Simon, B.M. (1980). Decomposition processes in the profundal region of Blelham Tarn and the Lund tubes. *Journal of Ecology* **68**, 493-512.

Jørgensen, B.B. (1977*a*). Bacterial sulphate reduction within reduced microniches of oxidized marine sediments. *Marine Biology* **41**, 7-17.

Jørgensen, B.B. (1977*b*). Distribution of colourless sulphur bacteria (*Beggiatoa* spp.) in a coastal marine sediment. *Marine Biology* **41**, 19-28.

Jørgensen, B.B. (1977*c*). The sulphur cycle of a coastal marine sediment (Limfjorden, Denmark). *Limnology and Oceanography* **22**, 814-832.

Jørgensen, B.B. and Fenchel, T. (1974). The sulphur cycle of a marine sediment model system. *Marine Biology* **24**, 189-201.

Kaplan, I.R., Emery, K.O. and Rittenberg, S.C. (1963). The distribution and isotopic abundance of sulphur in recent marine sediments off Southern California. *Geochimica et Cosmochimica Acta* **27**, 297-331.

Kellogg, W.W., Cadle, R.D., Allen, E.R., Lazrus, A.L. and Martell, E.A. (1972). The sulphur cycle. *Science* **175**, 587-596.
Kepkay, P.E., Cooke, R.C. and Novitsky, J.A. (1979). Microbial autotrophy: a primary source of organic carbon in marine sediments. *Science, N.Y.* **204**, 68-69.
Kepkay, P.E. and Novitsky, J.A. (1980). Microbial control of organic carbon in marine sediments: coupled chemotrophy and heterotrophy. *Marine Biology* **55**, 261-266.
King, G.M. and Wiebe, W.J. (1980). Tracer analysis of methanogenesis in salt marsh soils. *Applied and Environmental Microbiology* **39**, 877-881.
Kuenen, J.G. (1975). Colourless sulphur bacteria and their role in the sulphur cycle. *Plant and Soil* **43**, 49-76.
Laanbroek, H.J. and Pfennig, N. (1981). Oxidation of short-chain fatty acids by sulphate-reducing bacteria in freshwater and in marine sediments. *Archives of Microbiology* **128**, 330-335.
Latham, M.J. and Wolin, M.J. (1977). Fermentation of cellulose by *Ruminococcus flavefaciens* in the presence and absence of *Methanobacterium ruminantium*. *Applied and Environmental Microbiology* **34**, 297-301.
Liss, P.S. and Slater, P.G. (1974). Flux of gases across the air-sea interface. *Nature* **247**, 181-184.
Lovelock, J.E., Maggs, R.J. and Rasmussen, R.A. (1972). Atmospheric dimethyl sulphide and the natural sulphur cycle. *Nature* **237**, 452-453.
Mah, R.A., Ward, D.M., Baresi, L. and Glass, T.L. (1977). Biogenesis of methane. *Annual Review of Microbiology* **31**, 309-341.
Martens, C.S. and Berner, R.A. (1974). Methane production in the interstitial waters of sulphate-depleted marine sediments. *Science* **185**, 1167-1169.
Martens, C.S. and Berner, R.A. (1977). Interstitial water chemistry of anoxic Long Island Sound sediments. I. Dissolved gases. *Limnology and Oceanography* **22**, 10-25.
Martens, C.S. and Goldhaber, M.B. (1978). Early diagenesis in transitional sedimentary environments of the White Oak River estuary, North Carolina. *Limnology and Oceanography* **23**, 428-441.
McInerney, M.J., Bryant, M.P. and Pfennig, N. (1979). Anaerobic bacterium that degrades fatty acids in syntrophic association with methanogens. *Archives of Microbiology* **122**, 129-135.
Mechalas, B.J. (1974). Pathways and environmental requirements for biogenic gas production in the oceans. In "Natural Gases in Marine Sediments" (Ed. I.R. Kaplan), pp.12-25. Plenum Press, New York.
Miller, D., Brown, C.M., Pearson, T.H. and Stanley, S.O. (1979). Some biologically important low molecular weight organic acids in the sediments of Loch Eil. *Marine Biology* **50**, 375-383.
Molongoski, J.J. and Klug, M.J. (1980). Anaerobic metabolism of particulate organic matter in the sediments of a hypereutrophic lake. *Freshwater Biology* **10**, 507-518.
Mountfort, D.O. and Asher, R.A. (1979). Effect of inorganic sulphide on the growth and metabolism of *Methanosarcina barkeri* strain DM. *Applied and Environmental Microbiology* **37**, 670-675.
Mountfort, D.O., Asher, R.A., Mays, E.L. and Tiedje, J.M. (1980). Carbon and electron flow in mud and sandflat intertidal sediments

at Delaware Inlet, Nelson, New Zealand. *Applied and Environmental Microbiology* **39**, 686-694.

Nakai, N. and Jensen, M.L. (1964). The kinetic isotope effect in the bacterial reduction and oxidation of sulphur. *Geochimica et Cosmochimica Acta* **28**, 1893-1912.

Nedwell, D.B. and Abram, J.W. (1978). Bacterial sulphate reduction in relation to sulphur geochemistry in two contrasting areas of saltmarsh sediment. *Estuarine and Coastal Marine Science* **6**, 341-351.

Nedwell, D.B. and Abram, J.W. (1979). Relative influence of temperature and electron donor and electron acceptor concentrations on bacterial sulphate reduction in saltmarsh sediment. *Microbial Ecology* **5**, 67-72.

Nedwell, D.B. and Floodgate, G.D. (1972*a*). Temperature-induced changes in the formation of sulphide in a marine sediment. *Marine Biology* **14**, 18-24.

Nedwell, D.B. and Floodgate, G.D. (1972*b*). The effect of microbial activity upon the sedimentary sulphur cycle. *Marine Biology* **16**, 192-200.

Nissenbaum, A., Presley, B.J. and Kaplan, I.R. (1972). Early diagenesis in a reducing fjord, Saanich Inlet, British Columbia. I. Chemical and isotopic changes in major components of interstitial water. *Geochimica et Cosmochimica Acta* **36**, 1007-1027.

Nriagu, J. (1968). Sulphur metabolism and sedimentary environment: Lake Mendota, Wisconsin. *Limnology and Oceanography* **13**, 430-439.

Nriagu, J.O. and Coker, R.D. (1976). Emission of sulphur from Lake Ontario sediments. *Limnology and Oceanography* **21**, 485-489.

Oremland, R.S. and Taylor, B.F. (1978). Sulphate reduction and methanogenesis in marine sediments. *Geochimica et Cosmochimica Acta* **42**, 209-214.

Östlund, H.G. and Alexander, J. (1963). Oxidation rate of sulphide in seawater, a preliminary study. *Journal of Geophysical Research* **68**, 3995-3997.

Peck, H.D. (1959). The ATP-dependent reduction of sulphate with hydrogen in extracts of *Desulfovibrio desulfuricans*. *Proceedings of the National Academy of Science* **45**, 701-708.

Pfennig, N. (1975). The phototrophic bacteria and their role in the sulphur cycle. *Plant and Soil* **43**, 1-16.

Postgate, J.R. (1979). The Sulphate-Reducing Bacteria. Cambridge University Press.

Ramm, A.E. and Bella, D.A. (1974). Sulphide production in anaerobic microcosms. *Limnology and Oceanography* **19**, 110-118.

Reeburgh, W.S. (1980). Anaerobic methane oxidation: rate depth distributions in Skan Bay sediments. *Earth and Planetary Science Letters* **47**, 345-352.

Reeburgh, W.S. and Heggie, D.T. (1977). Microbial methane consumption reactions and their effect on methane distributions in freshwater and marine environments. *Limnology and Oceanography* **22**, 1-9.

Rees, C.E. (1973). A steady state model for sulphur isotope fractionations in bacterial reduction processes. *Geochimica et Cosmochimica Acta* **37**, 1141-1162.

Revsbech, N.P., Sørensen, J. and Blackburn, T.H. (1980). Distribution of oxygen in marine sediments measured with microelectrodes. *Limnology and Oceanography* **25**, 403-411.

Sørensen, J., Jørgensen, B.B. and Revsbech, N.P. (1979). A comparison of oxygen, nitrate, and sulphate respiration in coastal marine sediment. *Microbial Ecology* **5**, 105-115.
Sorokin, Y.I. (1962). Experimental investigations of bacterial sulphate reduction in the Black Sea using ^{35}S. *Microbiology* **31**, 329-335.
Sorokin, Y.I. (1964). On the primary production and bacterial activities in the Black Sea. *Journal du Conseil Permanent International pour Exploration de la Mer* **29**, 41-60.
Sorokin, Y.I. (1970). Interrelations between sulphur and carbon turnover in meromictic lakes. *Archives of Hydrobiology* **66**, 391-446.
Sorokin, Y.I. (1972). The bacterial population and processes of hydrogen sulphide oxidation in the Black Sea. *Journal du Conseil Permanent International pour Exploration de la Mer* **34**, 423-454.
Stanley, S.O., Pearson, T.H. and Brown, C.M. (1978). Marine microbial ecosystems and the degradation of organic pollutants. In "The Oil Industry and Microbial Ecosystems" (Eds. B. Chater and H. Somerville), pp.60-79. Heydon, London.
Strayer, R.F. and Tiedje, J.M. (1978). Kinetic parameters of the conversion of methane precursors to methane in a hypereutrophic lake sediment. *Applied and Environmental Microbiology* **36**, 330-340.
Tewes, F.J. and Thauer, R.K. (1980). Regulation of ATP-synthesis in glucose fermenting bacteria involved in interspecies hydrogen transfer. In "Anaerobes and Anaerobic Infections" (Eds. G. Gottschalk, N. Pfennig and H. Werner), pp.97-104. Gustav Fischer Verlag, Stuttgart.
Tezuka, Y. (1966). A commensalism between the sulphate reducing bacterium *Desulfovibrio desulfuricans* and other heterotrophic bacteria. *Botanical Magazine, Tokyo* **79**, 174-178.
Thauer, R.K., Jungermann, K. and Decker, K. (1977). Energy conservation in chemotrophic anaerobic bacteria. *Bacteriological Reviews* **41**, 100-180.
Thorstensen, D.C. (1970). Equilibrium distribution of small organic molecules in natural waters. *Geochimica et Cosmochimica Acta* **34**, 745-770.
Trudinger, P.A. (1969). Assimilatory and dissimilatory metabolism of inorganic sulphur compounds by microorganisms. *Advances in Microbial Physiology* **3**, 111-158.
Tsou, J.L., Hammond, D. and Horowitz, R. (1973). Interstitial water studies, Leg 15. Study of CO_2 released from stored deep sea sediments. In "Initial Reports of the Deep Sea Drilling Project" Vol.XX, pp.851-863. (Eds. B.C. Heezen *et al.*) US Government Printing Office, Washington D.C.
Tuttle, J.H. and Jannasch, H.W. (1973). Sulphide and thiosulphate-oxidizing bacteria in anoxic marine basins. *Marine Biology* **20**, 64-71.
Vosjan, J.H. (1975). Ecological and physiological aspects of bacterial sulphate reduction in the Wadden Sea (In Dutch). Thesis, University of Groningen.
Wellinger, A. and Wuhrmann, K. (1977). Influence of sulphide compounds on the metabolism of *Methanobacterium* strain AZ. *Archives of Microbiology* **115**, 13-17.
Widdel, F. (1980). Anaerober Abbau von Fettsauren und Benzoesaure

durch neu isolierte Arten Sulfat-reduzierender Bakterien. Doctoral thesis, University of Gottingen, West Germany.

Widdel, F. and Pfennig, N. (1977). A new anaerobic, sporing, acetate-oxidizing, sulphate-reducing bacterium, *Desulfotomaculum* (emend) *acetoxidans*. *Archives of Microbiology* **112**, 119-122.

Winfrey, M.R. and Zeikus, J.G. (1977). Effect of sulphate on carbon and electron flow during microbial methanogenesis in freshwater sediments. *Applied and Environmental Microbiology* **33**, 275-281.

Winfrey, M.R., Nelson, D.R., Klevickis, S.C. and Zeikus, J.G. (1977). Association of hydrogen metabolism with methanogenesis in Lake Mendota sediments. *Applied and Environmental Microbiology* **33**, 312-318.

Winfrey, M.R. and Zeikus, J.G. (1979). Anaerobic metabolism of immediate methane precursors in Lake Mendota. *Applied and Environmental Microbiology* **37**, 244-253.

Wolfe, R.S. (1971). Microbial formation of methane. *Advances in Microbial Physiology* **6**, 107-146.

Wolfe, R.S. and Higgins, I.J. (1979). Microbial biochemistry of methane - a study in contrasts. Part 1. Methanogenesis. *Microbial Biochemistry* **21**, 270-300.

Wolin, M.J. (1976). Interactions between H_2-producing and methane-producing species. In "Microbial Formation and Utilization of Gases" (Eds. H.G. Schlegal, G. Gottschalk and N. Pfennig), pp.141-150. E. Goltze K.G., Gottingen.

Zeikus, J.G. (1977). The biology of methanogenic bacteria. *Bacteriological Reviews* **41**, 514-541.

Zinder, S.H. and Brock, T.D. (1978). Methane, carbon dioxide, and hydrogen sulphide production from the terminal methiol group of methionine by anaerobic lake sediments. *Applied and Environmental Microbiology* **35**, 344-352.

Chapter 5

ACTIVITIES OF AEROBIC AND ANAEROBIC BACTERIA IN LAKE SEDIMENTS AND THEIR EFFECT ON THE WATER COLUMN

J. GWYNFRYN JONES

Freshwater Biological Association, The Ferry House, Ambleside, Cumbria, UK

Introduction

Considering its importance as a site of activity, it is surprising that so little has been written about the sediment as a habitat for microorganisms. Apart from the volume edited by Golterman [1977] and those written by Golterman [1975], and Fenchel and Blackburn [1979] modern texts rarely devote more than a few paragraphs to the microbiology of sediments. The plankton, on the other hand has been the subject of intense interest with many text-books and, more recently, whole journals devoted exclusively to its study. The bacterial population of the plankton is sparse (three to four orders of magnitude smaller than that of the benthos) and the extent of its involvement in carbon mineralization and associated geochemical cycles is often a fraction of that of the benthic bacteria. One of the purposes of this chapter is to illustrate the importance of the organisms in the sediment, and the extent to which they influence the overlying water. This volume contains a healthy bias towards marine and estuarine environments and therefore this particular contribution might also be seen as a small injection of fresh water.

Decomposition processes in lakes have been studied largely by comparison of mineralization with net primary production. Sediment respiration has often been taken as a measure of decomposition [Hargrave, 1969; Wetzel *et al.*, 1972; Jones, 1976] and has been shown to be related to lake trophic status [Ohle, 1956; Edmondson, 1966] and sediment particle size [Hargrave, 1972; Jones, 1980]. A few studies have attempted to determine the role of processes other than aerobic respiration (nitrification, oxidation of sulphur and iron) in oxygen consumption. Burns and Ross [1972] attributed 12% of the oxygen loss in Lake Erie to such inorganic processes, whereas Hall *et al*. [1978] calculated that nitrification alone accounted for 25% of the oxygen consumption in Grasmere, English Lake District. It is clear that although oxygen consumption may overestimate aerobic respiration, measurement of this single process of mineralization will lead to a serious underestimate of carbon turnover. Recent studies [Rudd and Hamilton, 1978; Barber and Ensign, 1979; Robertson, 1979; Fallon *et al*., 1980] have demonstrated the importance of methanogenesis, but very few investigations have attempted to determine the relative contributions of the various anaerobic decomposition processes in carbon mineralization in freshwater systems. This may reflect a recognition that the interactions of anaerobic bacteria in aquatic sediments are extremely complex [Abram and Nedwell, 1978; Cappenberg and Jongejan, 1978; Winfrey and Zeikus, 1977]. In a preliminary analysis of the profundal zone of Blelham Tarn, Jones and Simon [1980*a*] concluded that aerobic respiration accounted for approximately 42%, denitrification 17%, SO_4^{2-} reduction 2% and methanogenesis was equivalent to 25% of CO_2 accumulation. In contrast, detailed studies of sublittoral marine and estuarine sediments [Fenchel and Jørgensen, 1977; Fenchel, 1978] have shown SO_4^{2-} reduction to be the most important anaerobic process (equivalent in carbon terms, to 60% of aerobic respiration) whereas denitrification accounted for only a few percent and methanogenesis was considered to be negligible.

Almost all freshwater sediment microbiology has been confined to studies of the profundal (deep water) sediments, but recently [Jones, 1980] distinct differences have been observed between these and littoral (shallow water) sediments. This chapter will therefore include details of further investigations of these differences, and will also attempt to emphasize the main differences between freshwater and marine sediments. Some space will also be devoted to speculation on areas of ignorance and uncertainty.

The Lake Habitat

The lakes described and the results presented in this chapter are based on findings in the English Lake District, where the water is relatively soft and poorly buffered. There is sufficient evidence in the literature, however, to suggest that the processes observed are fairly typical of lakes in the temperate zone. Space does not permit a detailed discussion of the classification of lakes such as that given by Hutchinson [1957] or Odum [1971] but a few generalizations might be made which are relevant to the discussion on sediment microbiology. This is also a suitable point at which to define the few limnological terms which will be used in this chapter. For the purposes of the sediment microbiologist, lakes may be classified as deep or shallow. The deep lakes tend to be oligotrophic (nutrient poor) while the shallow lakes are more likely to be eutrophic (nutrient rich). The trophic status of a lake has two components, the cultural (that is the nutrient status related to enrichment from the catchment and man's activities) and the morphometric (this is related to the size and shape of the lake basin). In other words, a large deep lake is less likely to become deoxygenated and eutrophic than a shallow small one [Charlton, 1980]. This is simply because of the dilution of incoming nutrients and the volume of water which has to be deoxygenated largely through the activity of the organisms in the sediment. This chapter will be concerned to a large degree, with the shallow eutrophic lakes, because of the greater variety and intensity of microbial activity which is found in them. Some comparisons will, however, be drawn with more oligotrophic waters.

Most lakes in the temperate zone undergo a cycle of thermal stratification and destratification and this, more than any other, is the prime factor controlling bacterial activities. In spring the water column of the lake is isothermal and well oxygenated (Fig. 1). The main difference between the sediments of the littoral (shallow water) and profundal (deep water) zone is that the Eh gradient in the former is deeper in the sediment and not as steep, largely reflecting the greater turbulence at that site. As summer progresses the surface water warms until the lake becomes stratified into three distinct zones, the epilimnion (the warm, surface water), the metalimnion (the zone of temperature change - this contains the thermocline, the zone of steepest temperature gradient) and the hypolimnion (the cool, deep water). The temperature, and therefore density gradient of the water between the epilimnion and the hypolimnion ensures that the latter is effectively isolated and may be treated, approximately, as a closed system. The approximations and corrections which have to be made if such an assumption is made are dealt with later. The microbial processes which then take place in the hypolimnion are related to the quantity and quality of available organic carbon and the concentration of available electron acceptors. Aerobic respiration results in depletion of O_2 in the water column and other electron acceptors are then used in turn, thus creating more reducing conditions in the hypolimnion. The potential electron acceptors, their reduced products and approximate Eh ranges are summarized in Fig. 2, the ordinate of which could also have been represented as time, running from the onset of thermal stratification at +600 mV, to the end of summer at -300 mV. Thus the stratified eutrophic lake resembles Fig. 3 at the end of the summer. The hypolimnion is depleted of O_2 and the Eh gradient in the sediments is steeper. The Eh gradient in the profundal zone has migrated upwards until conditions at the sediment water interface become reducing. The net effect is a release of reduced end products into the

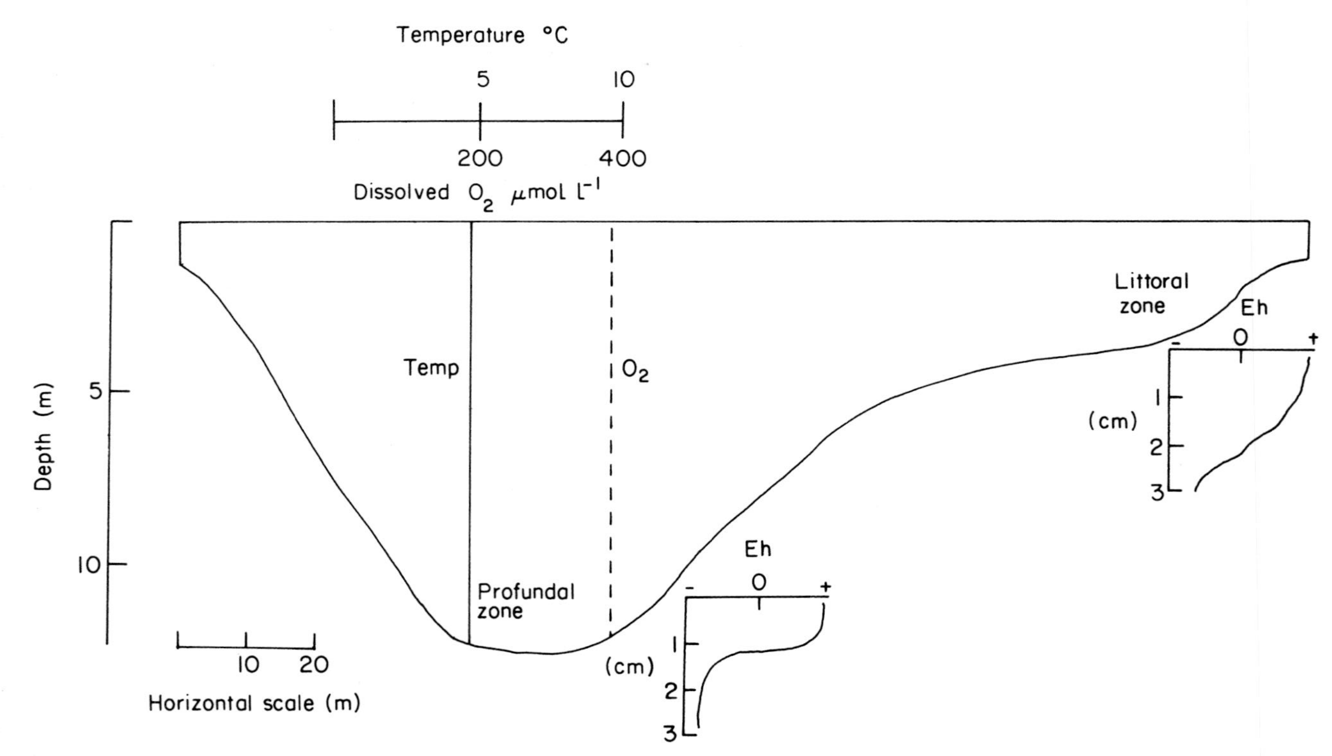

Fig. 1 Distribution of temperature, dissolved O_2 and sediment Eh with depth in a temperate shallow lake (winter).

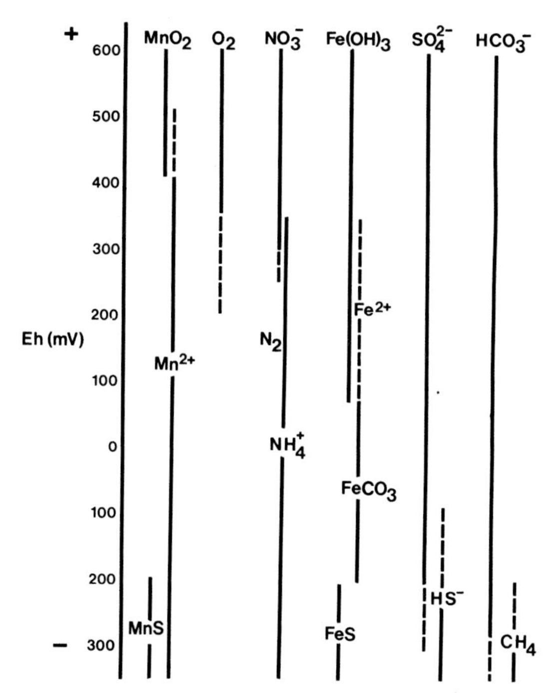

Fig. 2 Approximate Eh ranges of potential electron acceptors and their products in freshwater.

profundal water, and approximate values for the concentrations observed are shown in Table 1, as are the mean winter values for the whole water column. The values are markedly different from those obtained in marine systems, particularly with regard to the concentrations of potential electron acceptors. Seawater concentrations of SO_4^{2-} and NO_3^- are, on average, 28000 and 1 to 4 μmol l^{-1} hence the importance of sulphate reduction in marine decomposition [Fenchel and Blackburn, 1979; Jørgensen, 1980].

The main features of the freshwater hypolimnion in late summer are the absence of O_2 and NO_3^-, and considerable enrichment in NH_4^+, S^{2-}, Fe^{2+}, Mn^{2+}, P_i, CO_2 and CH_4. Large quantities of phosphorus are chemically bound to iron when the latter is in the oxidized state, and the P_i is released as the ferric iron is reduced to its soluble Fe^{2+} form. We owe much of our understanding of these processes to the classic papers of Mortimer [1941, 1942], work which appears to have been largely ignored by microbiologists, judging by the number of times that experiments similar to his have been repeated. Anyone interested in sediment-water interactions would benefit enormously from reading these papers. Although not necessarily described as such, the effects on the water column of aerobic respiration, nitrate, iron and

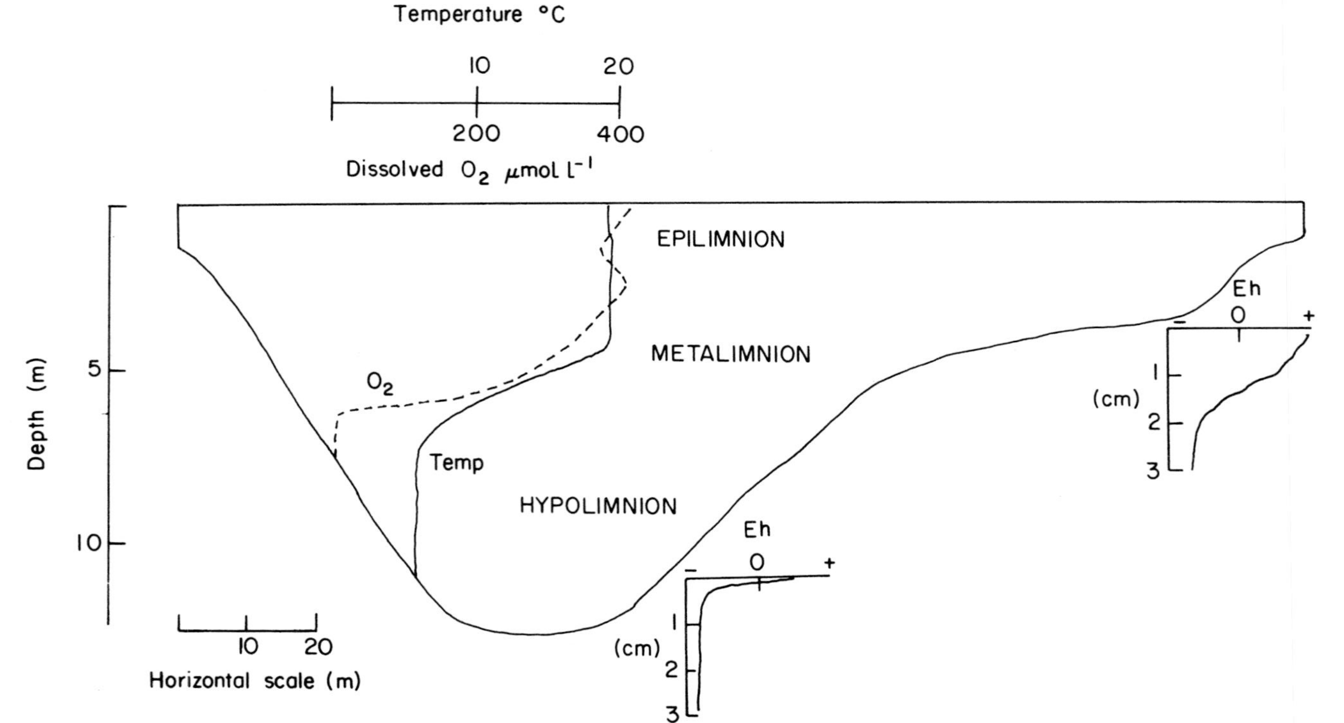

Fig. 3 Distribution of temperature, dissolved O_2 and sediment Eh with depth in a stratified eutrophic lake (summer).

TABLE 1

Concentration of nutrients in the water column of a temperate eutrophic lake (English Lake District)

	$\mu mol\ l^{-1}$		
	Winter (whole column)	Summer Epilimnion	Hypolimnion*
O_2	390	340	0
NO_3^-	66	50	< 1
NH_4^+	4	4	240
SO_4^{2-}	150	150	130
S^{2-}	nd	nd	18
Fe†	2.7	4	200
Mn†	0.5	1	45
P ‡	0.35	< 0.05	20
CO_2	500	250	1900
CH_4	nd	nd	580

* Values reported for the hypolimnion were obtained 1 m above the sediment water interface.
† Values given are for total iron and manganese although this will be composed largely of the ferrous and manganous ions in the anoxic hypolimnion in summer.
‡ Soluble reactive phosphorus as determined by reaction with molybdate.

sulphate reduction in the sediment are presented in such a way as to leave no doubt about the importance of microbial processes in controlling the Eh of the system and the chemistry of the overlying water.

In summary, the activity of bacteria in sediments is a major component in the production of anaerobic, reducing conditions in hypolimnion. This in turn enriches the hypolimnion in nutrients which stimulate bacterial production in the water column [Jones, 1977*a*]. The littoral sediments, on the other hand, are in a zone which is oxygenated, warmer in summer, and where the overlying water (containing other electron acceptors in addition to oxygen) is constantly replenished by inflowing rivers and streams. The net effect of these differences on geochemical cycling in the two zones is discussed later in the chapter.

General Remarks: Bacteria in Sediments, How Many and Where are They?

Almost without exception, research into the microbiology of sediments has been confined to the investigation of the profundal zone. Measurements of microbial biomass and activity (chlorophyll *a* concentration, direct counts of bacteria, ATP concentration, heterotrophic potential - the V_{max} for single substrate metabolism and electron transport system activity) have been used to rank lakes and compare these rankings with their apparent trophic status [Gillespie, 1976; Spencer, 1978; Jones *et al.*, 1979]. Some of the results obtained from 16 lakes in the English Lake District are shown in Fig. 4. There is a clear correlation between the increasing bacterial population and electron transport system activity in the sediments on the one hand, and the degree to which the hypolimnion of the lake becomes deoxygenated (Fig. 4*a*) on the other.

While such studies are useful for comparative purposes they provide little information on the variability of the sediment microflora and its activity within any given lake. Yet marked differences between littoral and profundal sediments exist and these should be taken into consideration when results are scaled up to provide "whole-lake" budgets. The fungal populations of marginal and deep water zones have been shown to differ [Willoughby, 1961], with members of the Saprolegniales being more active in the former [Willoughby, 1965; Dick, 1971]. On the other hand, Willoughby [1974] concluded that whereas some species were merely washed in, others were metabolically active in the deep water sediments. Until recently the only comparable study on bacteria was that of Willoughby [1969] who demonstrated that the distribution of actinomycetes differed from that of some lower fungi, in that numbers were highest in the deep water zone. Godinho-Orlandi and Jones [1981*a*,*b*] also observed larger numbers of a variety of other filamentous bacteria in the profundal zones of oligotrophic, mesotrophic and eutrophic lakes. Although few in terms of total numbers, these filamentous bacteria may form a significant component of the biomass of the sediment microflora. Jones [1980], in a more general study of oxidized surface sediments, showed that bacterial numbers (direct counts) biomass (ATP) and activity (electron transport system activity, CO_2 evolution and [^{14}C] glucose mineralization) were consistently higher in the profundal zone of mesotrophic and eutrophic lakes. This appeared to be due to the great availability of carbon and nitrogen in the profundal sediments (Table 2) although particle size distribution [Hargrave, 1972] and grazing by protozoa and larger invertebrates may also have contributed to the distribution observed [Fenchel, 1972; Fenchel and Jørgensen, 1977; Hargrave, 1970, 1976]. Although it is now possible

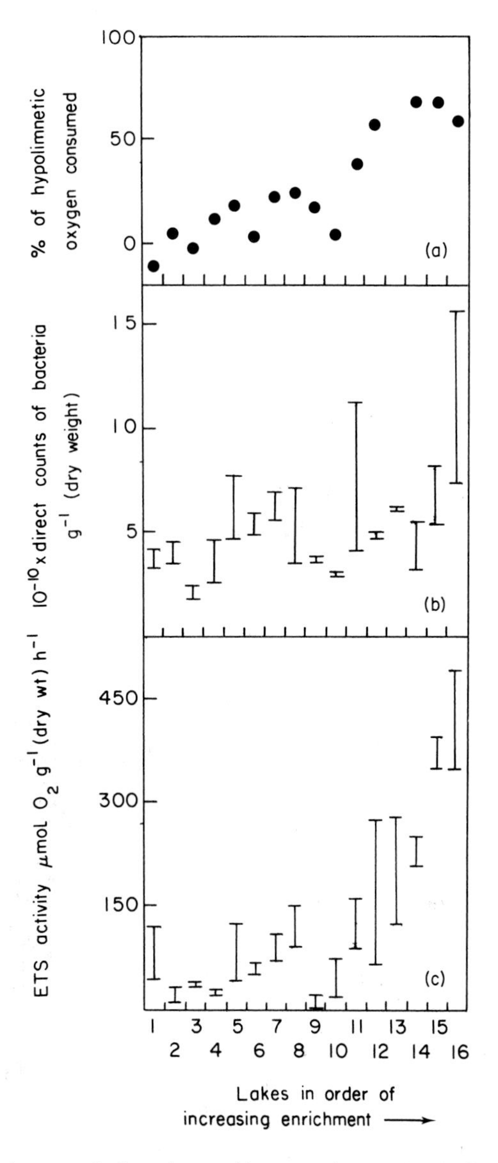

Fig. 4 Microbial activity in sediments in 16 Cumbrian lakes arranged in order of increasing enrichment. **(a)** Percentage of the hypolimnetic oxygen which is consumed during summer stratification. **(b)** Epifluorescence direct counts of benthic bacteria. Samples were taken from the profundal zone of each lake and the bars represent the range obtained during summer stratification. **(c)** Sediment electron transport system activity of profundal samples; bars represent the range obtained during summer stratification. Data from Jones *et al.* [1979].

TABLE 2

*Some differences in the physico-chemical and microbiological characteristics of littoral and profundal sediments in a stratified eutrophic lake**

	Sediment	
	Littoral	Profundal
Temperature, °C	14.5 (3.1)	6.8 (0.6)
Carbon content, mg g^{-1}	135 (16)	181 (0.6)
Nitrogen content, mg g^{-1}	13.2 (1.0)	21.8 (3.8)
Depth of Eh + 250 mV, mm	15 - 38	0 - 12
Dry weight, % (w/w)	14.2	5.2
Total carbohydrate, mg l^{-1}	9.8	39.0
Protein, mg l^{-1}	0.44 (0.17)	3.57 (2.78)
Amino nitrogen, μmol l^{-1} glycine equiv.	6.6	38.0
10^{-10} x Direct counts of bacteria g^{-1}	2.8 ± 0.3	11.9 ± 1.2
ETS activity (μmol O_2 g^{-1} h^{-1})	119	572
CO_2 evolution (μmol g^{-1} d^{-1})	4.5 (0.3)	28.2 (5.1)
ATP content (nmol g^{-1})	36.3	110.5

* Data from Jones [1980] and Jones and Simon [1981].

( ) indicates standard deviations, ± indicates 95% confidence limits

to provide some estimate of the microbiological variability which might be encountered in sediments within a lake, and to compare this both with variability at a single site and that which occurs over the season [Jones and Simon, 1980*b*], such studies have been largely confined to the oxygenated surface sediments. Clearly, differences between littoral and profundal sediments (in degree of oxygenation and the depths of the Eh gradient) will contribute to the interaction of aerobic and anaerobic processes, and the relative importance of the anaerobic respirations. Research is just beginning on the horizontal zonation of sediment decomposition processes within lakes. Klug and his coworkers [Klug *et al.*, 1980] have shown a shift in metabolic pattern between littoral and profundal zones and Jones and Simon [1981] demonstrated marked differences particularly in inorganic nitrogen metabolism. These differences are discussed in more detail later in this chapter.

Before turning to the details of bacterial processes in freshwater sediments, one further observation might be made on the general distribution of bacteria in relation to Eh gradients within the freshwater systems. As has been shown earlier, such gradients gradually move upwards and eventually enter the water column as the sediments of eutrophic lakes become anoxic. Microbial communities in sediments have been sampled at such broad depth intervals [e.g. Johnston and Cross, 1976; Vosjan and Olanczuk-Neyman, 1977; Wieser and Zech, 1976] that the zone of Eh discontinuity was often encompassed in a single sample. The picture which has emerged has been one of a steadily decreasing activity and population size with depth. The importance of Eh gradients as sites of high metabolic activity has, however, been recognized [Fenchel and Jørgensen, 1977], and it is possible to demonstrate the effect of such gradients in freshwater systems on both a large (metre) and small (millimetre) scale. If transect sediment samples are taken across a stratified eutrophic lake a peak of microbial biomass and activity may be observed in the surface sediment associated with the metalimnion (Fig. 5). This is the zone where the Eh of the surface sediment drops and the microbial populations may be bathed (due to internal water movements) alternatively in the nutrient rich water of the hypolimnion and oxygenated water from the epilimnion. It is hardly surprising that large populations of filamentous sulphur bacteria and, if there is sufficient light penetration, cyanobacteria are found in this zone. On a smaller scale,

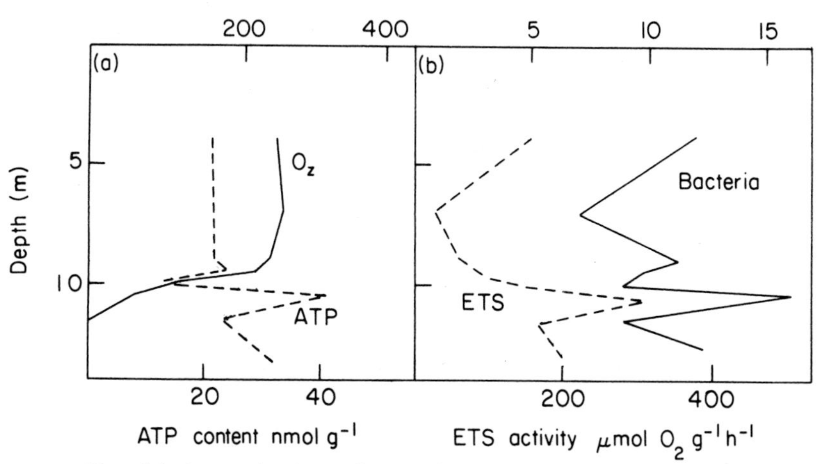

Fig. 5 Microbial populations in surface sediment in a transect across a stratified eutrophic lake. **(a)** Dissolved oxygen (——) and ATP content (-----). **(b)** Epifluorescence direct counts of bacteria (——) and electron transport system activity (----). Data from Jones [1979*a*].

sharp gradients of Eh are found over a depth of a few millimetres within the sediment. Once again, peaks of electron transport system activity may be observed on the gradient (Fig. 6) and these peaks move upwards with the

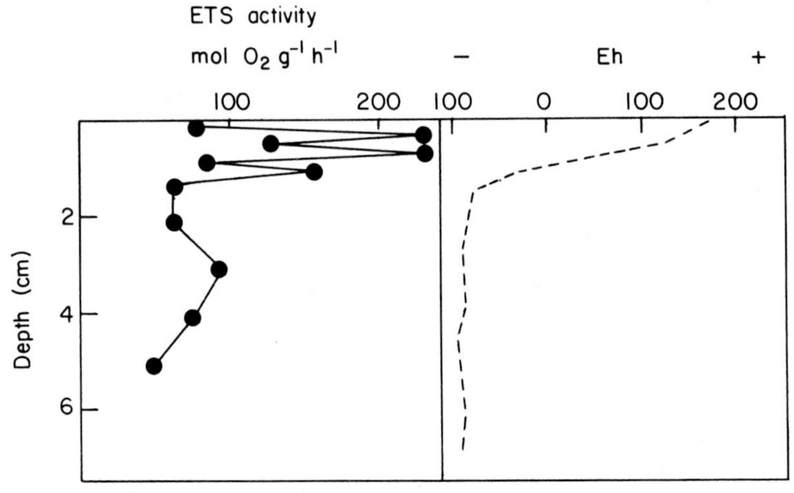

Fig. 6 Distribution of Eh and electron transport system activity with depth in a profundal sediment core. Data from Jones [1979*a*].

Eh gradient as summer progresses [Jones, 1979*a*]. Such measurements (over a few millimetres) of Eh are still on an enormous scale compared with the microniches which the bacteria may inhabit, and should not be treated as an infallible guide to potential bacterial activity. Revsbech *et al.* [1980] have, through the use of micro-electrodes, demonstrated that O_2 disappears with depth in marine sediments long before a significant reduction in Eh is detected. Likewise, Jørgensen [1977] has shown that large particles can act as microniches for SO_4^{2-} reduction in aerobic sediments. If methane can be generated in microniches in the apparently aerobic open ocean [Oremland, 1979] then freshwater microbiologists might benefit from a re-examination of their long held beliefs about lake sediments.

Bacterial Activity in the Sediments - the Whole Lake Approach

In the section on the lake habitat it was stated that the hypolimnion could be treated as a closed system (Fig. 3). In such a system the net change in electron acceptor concentration and the accumulation of reduced end products should provide an estimate of the relative importance of decomposition processes. Before such calculations can be considered, however, certain factors must be taken into account. In the first place, the hypolimnion can not be considered to be a truly closed system, and corrections must be made for inputs and outputs. These are as follows:

Inputs (a) Sedimenting particulate material.
(b) Transport by turbulence of electron acceptors, particularly O_2 across the metalimnion.
Outputs (a) Gas bubbles (particularly CH_4 and N_2)
(b) Transport by turbulence of reduced end products, across the metalimnion.

In addition to the above, organisms themselves may result in net transport across the metalimnion, but such a mechanism does not appear to have received any attention to date. Of the mechanisms listed above, Input (a) and Output (a) may be measured directly with sediment and gas traps respectively. The transport by turbulence (units, mmol cm^{-2} s^{-1}) may be calculated from the product of the turbulence coefficient (units, cm^2 s^{-1}) and the concentration gradient (units, mmol cm^{-3} cm^{-1}) at the depth under consideration. Examples of such a calculation for O_2 input to the hypolimnion are given by Jones [1976]. Whereas the concentration gradient may be measured with relative ease, the calculation of turbulence (eddy conductivity) coefficients may be subject to several errors, particularly in the region of the metalimnion. Turbulence may be calculated from changes in conservative properties of the water column (e.g. temperature) and examples of such calculations are provided by Mortimer [1941] and McEwan [described by Hutchinson, 1941]. A useful discussion is provided by Smith [1975] and a practical consideration by Powell and Jassby [1974].

Given that reasonable estimates of the above factors can be obtained, then the remaining error in such calculations is associated with the measurement of net change in a dynamic system. Most of the reduced end products of the respirations or hydrogen transport (NH_4^+, Mn^{2+}, Fe^{2+}, S^{2-}, CH_4) may be reoxidized by microbiological or chemical agencies. Measurements of methane oxidation or nitrification, for example, must be made before quantitative estimates of methanogenesis and nitrate reduction in the hypolimnion can be obtained. The net changes may, however, provide useful information on the relative importance of the various anaerobic microbial processes if the study area excludes the zones of reoxidation.

Although there is ample evidence that processes in the sediment tend to govern changes in the hypolimnion of stratified eutrophic lakes, it is often useful to provide additional evidence of the extent of direct influence of the sediment on the overlying water column. A first approximation of this may be obtained using the method described by Edberg [1976]. By applying a solution of Fick's second law of diffusion, the logarithm of the change in concentration of any chosen determinant is related in a linear manner to the distance from the sediment. Within the distance from the sediment where this relationship holds, it may be assumed that the effect is due to sediment activity. An example is provided in Fig. 7, where the logarithm of the decrease in oxygen concentration is linear with distance from the sediment for 6 m. It might therefore be concluded that the major cause of deoxygenation of the water over this depth was sediment activity. Bearing in mind that active sediment exists at all depths in the hypolimnion, and the potential for lateral water transport, it becomes easier to appreciate the influence of the benthic microflora on the water column.

The remainder of this section is devoted to a selection of field data which have allowed some estimates of sediment microbial activity to be made.

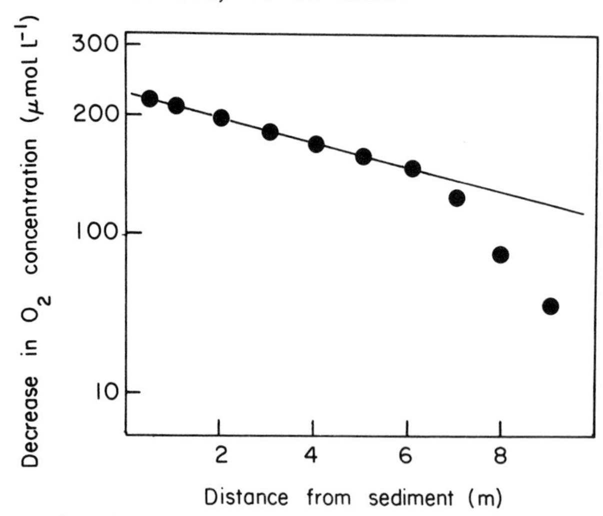

Fig. 7 Determination of the extent of the effect of sediment activity on the water column by measurement of decrease in O_2 concentration with distance from the sediment.

Aerobic Respiration

The aerobic respiration of bacteria causes rapid deoxygenation of the hypolimnion in summer (Fig. 8). From such data the hypolimnetic areal oxygen deficit may be calculated, and this has been used to compare lakes of different trophic status, although its relationship to lake productivity is not a simple one [Charlton, 1980]. A detailed examination of deoxygenation in Blelham Tarn (English Lake District) was made by Jones [1976] who, applying the corrections discussed above, obtained a deficit of approximately 15 mmol O_2 m^{-2} day^{-1}. This compared with a deficit for the eutrophic Esthwaite Water of

20 mmol m^{-2} day^{-1}, for Windermere N. Basin of 16 mmol m^{-2} day^{-1}, for the oligotrophic Ennerdale Water of 3 mmol m^{-2} day^{-1} [Mortimer, 1941] and for Lake Erie of 12 mmol m^{-2} day^{-1} [Burns and Ross, 1972]. Hutchinson [1957] related the rate of hypolimnetic O_2 loss to the type of lake and proposed ranges of 1.25 to 10 mmol m^{-2} day^{-1} for oligotrophic and 15 to 44 mmol m^{-2} day^{-1} for eutrophic lakes, but he also suggested that Mortimer's boundaries of < 8 and > 17 mmol m^{-2} day^{-2} for oligotrophic and eutrophic lakes might be more convenient.

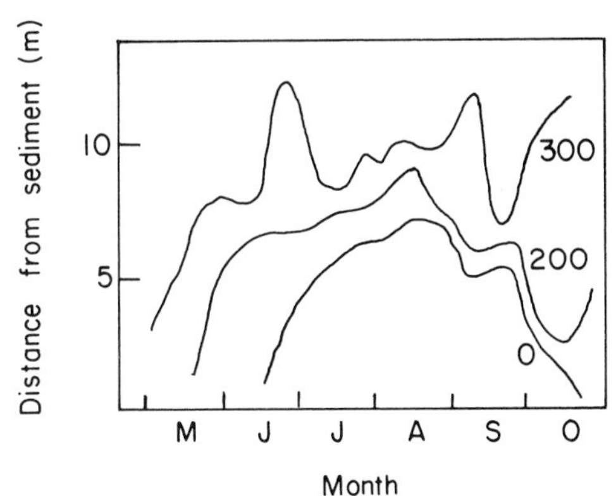

Fig. 8 Isopleths of dissolved oxygen concentration (μmol l^{-1}) in stratified eutrophic lake. Data from Jones and Simon [1980*a*].

By application of standard stoichiometric equations [Richards, 1965] or assumed RQ values, it was possible to calculate that deoxygenation of the hypolimnion would account for 52% of the primary production of Blelham Tarn and 86% of the input via sedimentation. Jones [1976] compared these values with those obtained on a variety of other lakes. Measurements based on O_2 alone, however, underestimate carbon turnover [Wetzel *et al.*, 1972] and this is particularly true in lakes where CO_2 accumulation is more rapid under anaerobic conditions [Burns and Ross, 1972; Hutchinson, 1957; Ohle, 1956]. Therefore the role of anaerobic microorganisms must also be considered.

Anaerobic Processes

After the removal of dissolved O_2 from the water, organic carbon decomposition continues through the agencies of NO_3^-, Fe^{3+} and SO_4^{2-} reduction, methanogenesis and anaerobic fermentations. The respirations occur in approximately the order in which they are listed because they are involved in electron transport phosphorylation and therefore redox linked reactions [see also Billen, this volume]. Fermentations on the other hand, which involve substrate level phosphorylation, may occur at any

Eh value in the absence of oxygen.

Theoretically the removal of oxygen should be followed closely by the reduction of manganese to the Mn^{2+} ion (Fig. 2) and this should precede Fe^{3+} reduction. Such a sequence has been observed in Esthwaite Water [Mortimer, 1971] and in Lake Vechten, The Netherlands [Verdouw and Dekkers, 1980] but unfortunately there appears to have been little consideration of the potential role of bacteria in this process. Although there is some evidence for the bacterial reduction of MnO_2 to Mn^{2+} [Woolfolk and Whiteley, 1962; Trimble and Ehrlich, 1970] and that the presence of MnO_2 inhibits Fe^{3+} reduction by bacteria [Munch and Ottow, 1977] it has yet to be shown that this is an energy yielding process.

The continuing accumulation of CO_2 in the hypolimnion under anaerobic conditions is accompanied by nitrate reduction, Fe^{3+} reduction, SO_4^{2-} reduction and methanogenesis and some of these events are illustrated in Fig. 9.

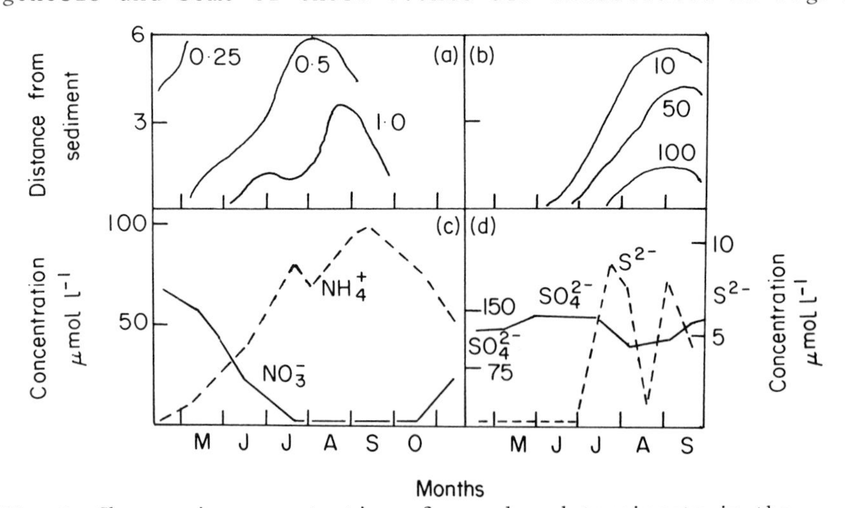

Fig. 9 Changes in concentration of some key determinants in the hypolimnion of a eutrophic lake during summer stratification. **(a)** Isopleths of carbon dioxide concentration (mmol l^{-1}). **(b)** Isopleths of dissolved CH_4 concentration (μmol l^{-1}). **(c)** Change in nitrate and ammonium concentration 1 m above the sediment. **(d)** Change in sulphate and sulphide concentration 1 m above the sediment. Data taken from Jones and Simon [1980*a*].

The CO_2 and CH_4 data are presented as depth-time diagrams with isopleths (contours) of equal concentration, illustrating the build up in their concentration from the sediment water surface. The inorganic nitrogen and sulphur data are presented as the change in concentration 1 m above the sediment as the summer progresses, as the presence of nitrification and sulphide deposition as FeS in the water column would have produced an extremely complicated depth-time diagram. The nitrate is completely removed from the profundal water by the end of July and this is followed by a period of accumulation of S^{2-} and CH_4. The build up of NH_4^+ in the water column stops temporarily when the NO_3^- concentration is zero, but then a further, relatively small increment of NH_4^+ appears. Presumably this is the result of anaerobic ammonification. The sources of reduced N and S and the fate of nitrate are discussed later. Nitrification may affect calculations of carbon turnover which are based on net removal of NO_3^-, but this is more likely to be a significant process in mesotrophic lakes [Hall *et al.*, 1978]. In oligotrophic lakes the NH_4^+ concentration in the hypolimnion is unlikely to reach levels which are high enough to support nitrifiers. In eutrophic lakes, on the other hand, the profundal zone deoxygenates so rapidly that there is insufficient time for a population of nitrifying bacteria to develop; one might therefore expect them to play a more significant role in the metalimnion. Similarly, errors may arise in the estimates of carbon mineralization derived from SO_4^{2-} reduction if these are not corrected for reoxidation of the sulphide. In early summer, before the profundal zone becomes completely reduced, the appearance of large populations of *Beggiatoa* spp. on the surface of the sediment presumably contributes to this reoxidation. This period may last for only a few days in a eutrophic lake, whereas the role of *Beggiatoa* has been shown to be of much greater significance in the sulphur and carbon cycles of marine sediments [Jørgensen, 1977]. Presumably chemical oxidation of S^{2-} in the neutral pH, oxygenated overlying water contributes to an underestimate of SO_4^{2-} reduction at certain times of the year but precipitation and resedimentation as FeS in late summer [Davison *et al.*, 1981] is more likely to affect calculations based on H_2S determinations. Such FeS precipitation could only occur where significant quantities of Fe^{3+} were being reduced, but this is usually the case in temperate eutrophic lakes. The build up of Fe^{2+} ions in the hypolimnion [examples are provided by Jones, 1978] as a result of the activity of the sediment bacteria is considerable (Table 1). Bacterial reduction of Fe^{3+} has been demonstrated [Ottow and Glathe, 1971; Woolfolk and Whiteley, 1962] and at least some of this may be attributed to nitrate reductase activity [Ottow and Munch, 1978; Balashova and Zavarzin, 1979; Thauer *et al.*, 1977]. Although there is no direct evidence of Fe^{3+} reduction linked to electron transport phosphorylation in bacteria, a few reports on the stimulation of growth and CO_2 production in the presence of Fe^{3+} have appeared [Takai and Kamura, 1966; Balashova and Zavarzin, 1979]. Unfortunately, it is impossible to obtain a quantitative relationship between the appearance of Fe^{2+} in the hypolimnion and any production of CO_2 which may be associated with it.

The appearance of CH_4 in the water column, and the sensitivity with which it may be determined provides a reliable estimate of the importance of this process. It has also been possible to determine losses due to turbulent exchange with the epilimnion and bacterial oxidation at any point, in the sediment or the water column, where O_2 and CH_4 gradients cross [Rudd and Hamilton, 1978].

In addition to the release of soluble components from the sediment, bacterial activity may result in the discharge of gaseous end products, particularly CH_4 and N_2, which has to be measured if reliable estimates of methanogenesis and denitrification are to be obtained. Some results obtained with gas traps placed in a eutrophic lake are shown in Fig. 10. The rate of release increases

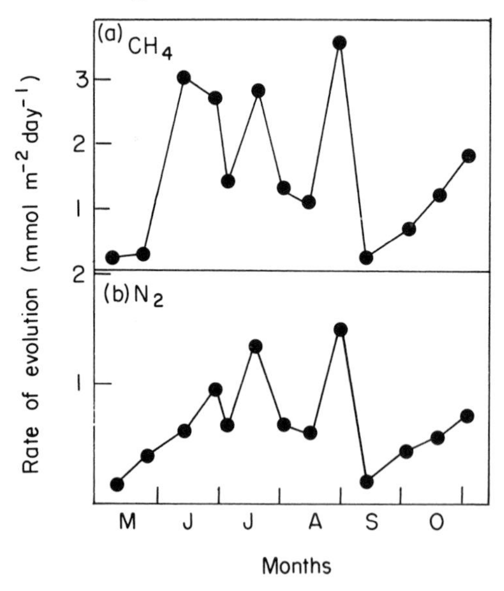

Fig. 10 Rate of evolution of **(a)** methane and **(b)** nitrogen in bubbles released from the profundal sediment during summer stratification in a eutrophic lake. Data from Jones and Simon [1980].

steadily in early summer and then fluctuates during the summer, depending on prevailing conditions and net C input to the sediment [Jones and Simon, 1981]. In the Cumbrian lakes where gas release has been measured, its production almost always drops dramatically in late summer. This has yet to be explained, but may well be associated with limitation in available C or H_2, more likely the latter. At autumn overturn there is a further release of gas, presumably associated with increased turbulence. Significant release of gas in winter is usually associated with ice cover and inverse stratification of the lake. The composition of the gas is remarkably constant for most of the season (35% N_2, 65% CH_4) and fairly similar to that collected by Chen *et al.* [1972] from the sediments of Lake Mendota (24 to 50% N_2, 45 to 75% CH_4) and by Snodgrass [1976] from the deeper water of Hamilton harbour (23% N_2, 70% CH_4). Howard *et al.* [1971] on the other hand, found that the composition of the gas released from the sediments of Lake Erie was 95% CH_4, 3% N_2 and 2% CO_2. The rate of methanogenesis in Blelham Tarn, the eutrophic lake from which much of the data presented here were obtained, is approximately 2.5 mmol m^{-2} day^{-1}. This is up to an order of magnitude lower than rates from many of the intensively studied lakes in North America [Hayward, 1968; Howard *et al.*, 1971; Molongoski and Klug, 1980; Rudd and Hamilton, 1978; Robertson, 1979; Strayer and Tiedje, 1978] and equivalent to much less of the carbon input (10 to 20%) than values reported by Robertson [1979], Rudd and Hamilton [1978], Molongoski and Klug [1980] and Fallon *et al.* [1980]. The rates were more comparable to the 0.7 mmol m^{-2} day^{-1} reported by Barber and Ensign [1979] for Lake Wingra, Wisconsin. Differences in patterns of gas release between littoral and profundal sediments are considered later in this chapter.

Although several workers have measured gas release from sediments, little consideration appears to have been given to errors inherent in the use of gas traps. Corrections are required for the solubility of CH_4 and, if the trap is in oxygenated water, oxidation by methylotrophs while the bubble is standing in the trap. In addition, CH_4 bubbles may strip N_2 from solution as they form in the sediment, as they travel through the water column and in the trap itself. In practice, corrections to the methane values are negligible but significant quantities of N_2 may enter the gaseous phase through physical processes rather than through the activities of denitrifiers. The values presented by Jones and Simon [1981] are corrected, as far as is possible, for these factors.

The Relative Importance of Decomposition Processes

It is possible to obtain an approximate estimate of the relative importance of the various bacterial decomposition processes which occur in the sediment from field observations such as those described above. An example of calculations based on such data is presented in Table 3. The values represent the sums of the net inputs, outputs and changes in the hypolimnion during summer stratification. It is immediately apparent that aerobic respiration, NO_3^- reduction and methanogenesis are quantitatively far more important than SO_4^{2-} reduction in carbon turnover in freshwater sediments. This is in marked contrast to the marine system where SO_4^{2-} reduction can be equivalent to 60% of aerobic respiration, denitrification may account for only a few percent and methanogenesis is considered to be unimportant [Fenchel and Jørgensen, 1977; Fenchel,

TABLE 3

Summary of decomposition processes in the profundal zone of a eutrophic lake during summer stratification. Results derived from field observations.

Standing crop of particulate material in the epilimnion (g atom m^{-2})		
Carbon	0.93 ± 0.18	
Nitrogen	0.11 ± 0.03	
Input to the hypolimnion as particulate material (g atom m^{-2})		
Carbon	6.3	
Nitrogen	0.73	
Output from the hypolimnion as gas bubbles (mol m^{-2})		
CH_4	0.28	(16%)*
N_2	0.124	(17%)
Net changes in the hypolimnetic water column (mol m^{-2})		
CO_2 accumulation	1.77	
O_2 uptake	0.98	(42%)
CH_4 accumulation	0.16	(9%)
NO_3^- consumption	0.174	(13%)
NH_4^+ accumulation	0.165	(12%)
S^{2-} accumulation	0.0139	(1.7%)

* These values represent the % contribution of each process to the production of CO_2, derived from the stoichiometric formulae of Richards [1965]. Methane values are converted to % equivalents of the total CO_2 produced. Data from Jones and Simon [1980*a*].

1978; Jørgensen, 1980]. Thus the concentrations of the various electron acceptors in these environments (Table 1) control the relative roles of the respiratory decomposition processes mediated by bacteria. The second conclusion to be drawn from Table 3 is that a significant proportion of the CO_2 which accumulates cannot be attributed to the processes listed. Both Fe^{3+} reduction (Eh dependent) and anaerobic fermentations (Eh independent) could contribute to this CO_2 imbalance, and this emphasizes our general ignorance of their importance in freshwater ecosystems.

The Organisms: the Sediment as a Refuge

Lest this chapter should become so engrossed in the chemistry of microbial processes that the organisms themselves are forgotten, this appears to be a suitable point at which to note that it is not only the end products of microbial metabolism which move between the sediment and the water column, but also the organisms themselves. The sediment may act as a winter refuge for organisms which have a planktonic existence in the summer. This is particularly true of bacteria which are found in the hypolimnion, such as *Leptothrix* sp. and *Ochrobium* sp. which are implicated in iron transformations [Jones, 1981] and a number of filamentous bacteria [Godinho-Orlandi and Jones, 1981*a,b*]. Many such organisms possess gas vacuoles which may assist in buoyancy regulation [Clark and Walsby, 1978*a,b*]. Larger eukaryotes such as ciliates [Finlay, 1981] possess the ability to swim actively and may move up from the sediment as anoxic conditions develop. Their return in winter is presumably to obtain an adequate supply of food. Conversely, dinoflagellates such as *Ceratium hirundinella* are to be found as resting stages overwintering in the sediment [Heaney and Talling, 1980].

Bacterial Activity in Sediments: Direct Measurement and Experimental Approach

The errors associated with the whole lake approach are apparent in the previous section of this chapter. Those associated with the experimental approach revolve largely around the problem of obtaining an undisturbed sample of sediment. The majority of the microbial processes to be studied are Eh related and therefore any disturbance of the steep gradients illustrated in Figs 1 and 3 will clearly affect the measurements made. For example, careless handling of a sediment core would result in the release of reduced chemical species into the overlying water. These possess a significant chemical oxygen demand which, in turn, would affect any estimate of aerobic respiration based on oxygen uptake. Some (relatively few) investigations have studied the activity of benthic microbes, and its effect on the water column, by using SCUBA divers to insert chambers into the sediment *in situ*. This procedure works well as long as the effect of reduced turbulence within the chamber is appreciated. It is particularly suitable for work in relatively shallow, clear water and has been used most extensively for measurements of benthic primary production in the littoral zone. It is less successful in the anoxic profundal zone, where darkness, the turbidity of the water and the soft flocculent nature of the sediment hamper diving operations. On many occasions therefore, it is necessary to remove a sample to the laboratory, usually in a corer. It is not the purpose of this chapter to discuss sampling apparatus, descriptions of which may be found in Collins *et al.* [1973] and in the bibliography by Elliott and Tullett [1978].

If disturbances of the sediment can be kept to a minimum, the advantages of working with samples in the laboratory are many. Apart from the ability to control experimental conditions it is also possible to obtain rapid comparisons of samples from different zones of the lake, and to determine the activity of bacteria as a function of depth in the sediment. The use of cores also provides better short term estimates of processes involving inter-species hydrogen transfer than those obtained with sediment slurries. This is discussed in more detail later. The addition of radioisotopes to cores allows many processes to be studied to a high degree of sensitivity, but perhaps more consideration should be given to the use of heavy isotopes, particularly ^{15}N, in field studies.

The examples of experimental measurements presented here have been chosen to illustrate the differences between littoral and profundal sediments and the interactions which may occur between bacterial populations.

Aerobic Respiration

Cores which enclose sediment and some of the overlying water have been used since Mortimer [1941, 1942] to obtain estimates of sediment oxygen uptake. The measured respiration rates have been related to lake trophic status [Ohle, 1956; Edmondson, 1966], sediment particle size [Hargrave, 1972] and the effect of total and mixed-layer depth on the benthic communities [Hargrave, 1975]. The measurements must be made in such a way that disturbance is kept to a minimum and light (which may stimulate benthic photosynthesis) excluded. With care, the results obtained may be used to simulate hypolimnetic deoxygenation, if allowances are made for lake morphometry and the change in respiration rate with O_2 concentration [Jones, 1976]. Oxygen uptake decreases with O_2 concentration as summer progresses (Fig. 11a) and reappears during the autumn overturn. The relationship between uptake and O_2 concentration approximates to a rectangular hyperbola (Fig. 11b) from which an apparent V_{max} and K_m may be calculated and used to model hypolimnetic deoxygenation. Others have successfully used power functions of O_2 concentration for similar calculations [Edwards and Rolley, 1965; Edberg and Hofsten, 1973]. The mean respiration rate for Blelham Tarn sediment is 21 mmol O_2 m^{-2} day^{-1} which is in reasonable agreement with the values cited earlier for comparable lakes. These were obtained with relatively undisturbed cores incubated for short time periods. If the water overlying the sediment is stirred then oxygen demand is increased [Edwards and Rolley, 1965] and if unrealistic stirring rates are used then oxygen consumption of lake sediments may be overestimated by more than an order of magnitude [James, 1974]. This overestimate will be due in part to the increase in chemical oxygen demand mentioned earlier, corrections for which may be obtained from formaldehyde poisoned cores [Jones, 1977*b*].

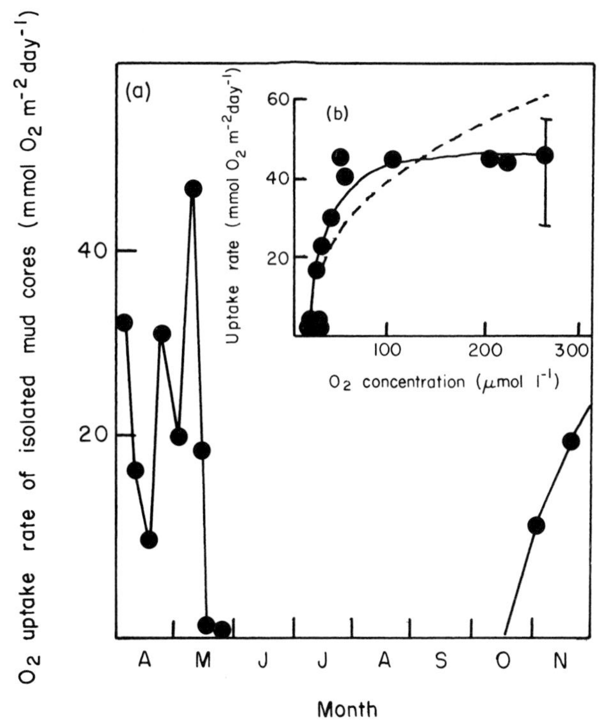

Fig. 11 Experimental measurements of O_2 uptake in isolated sediment cores from a stratified eutrophic lake. **(a)** Seasonal changes in aerobic respiration rate. **(b)** The effect of dissolved O_2 concentration on uptake in a sealed Jenkin core. Data from Jones [1976].

In addition to this, organisms which were previously respiring at concentrations close to their K_m values for O_2 (bearing in mind the microelectrode results of Revsbech *et al.*, 1980) will be able to consume oxygen at rates closer to their V_{max}. It is impossible to correct for such a change at present.

Comparative estimates of respiratory activity in the oxygenated surface sediments may also be obtained from measurements of electron transport system activity. These illustrate the greater potential for activity in the profundal zone [Jones, 1980], but allowance must also be made for the fact that not only is the littoral zone warmer in the summer but it is also in contact with oxygenated water throughout the period [Jones and Simon, 1981].

Anaerobic Processes

Below the zone of aerobic respiration, Eh determinations indicate that the zones of NO_3^- reduction, SO_4^{2-} reduction and methanogenesis should be arranged in that order with increasing depth of sediment [Jørgensen, 1980]. The distribution of fermentative organisms is not necessarily related to Eh but evidence exists of a change from a tri-carboxylic acid cycle-based metabolism to a fermentative one with increasing depth and decreasing electrode potential [Jones, 1979*a*]. As summer progresses, an upward movement in the Eh gradient may be observed, and in eutrophic systems the inorganic electron acceptors, NO_3^- and SO_4^{2-}, become limiting. The surface sediment becomes the major site of fermentative and methanogenic activity [Molongoski and Klug, 1980].

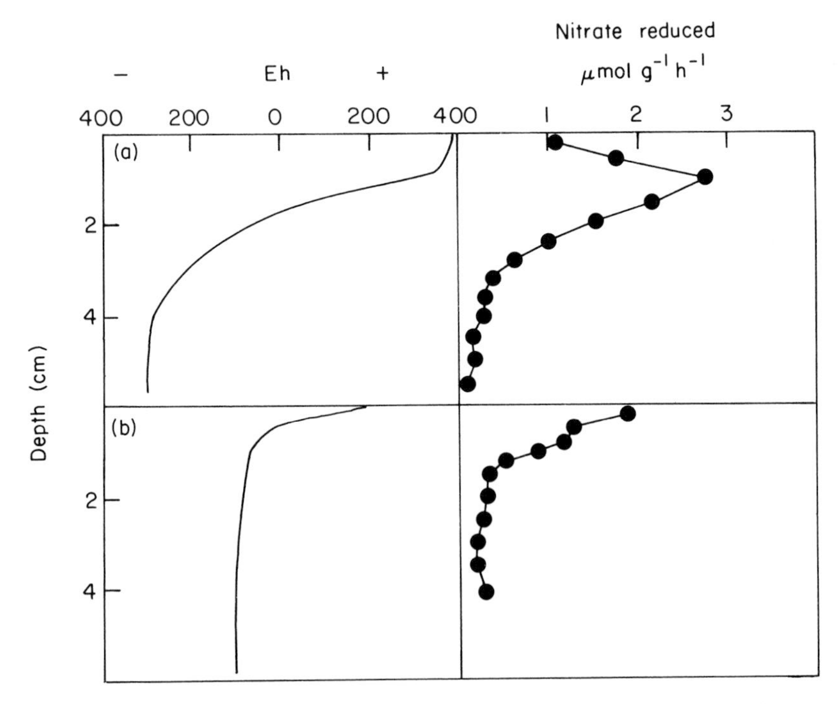

Fig. 12 Distribution of Eh and nitrate reductase with depth in **(a)** littoral and **(b)** profundal sediment of a stratified eutrophic lake. Data from Jones [1979*b*].

An example of the effect of the Eh profile on the depth of the nitrate reduction zone (in this case nitrate reductase activity) is given in Fig. 12. Sub-surface peaks of denitrification have also been observed in marine sediments by Sørensen [1978*a*,*b*]. While the sediment is oxidized at the surface, the peak of nitrate reductase activity is observed at a mean Eh of +210 mV. This is slightly higher than the Eh at which denitrification and nitrate poising of sediments has been reported by

Kessel [1978], Johnston *et al.* [1974] and Graetz *et al.* [1973]. The role of dissimilatory nitrate reductase increases in importance with depth [Jones, 1979*b*] and significant differences in the quantities of NH_4^+ which accumulate in littoral and profundal sediments are observed (Fig. 13). The relative importance of denitrification and NO_3^- fermentation to NH_4^+ at the two sites is discussed later. Using the acetylene inhibition technique described by Sørensen [1978*a*,*b*], denitrification rates of 0.8 to 2.6 mmol m^{-2} day^{-1} have been obtained [Jones and Simon, 1981].

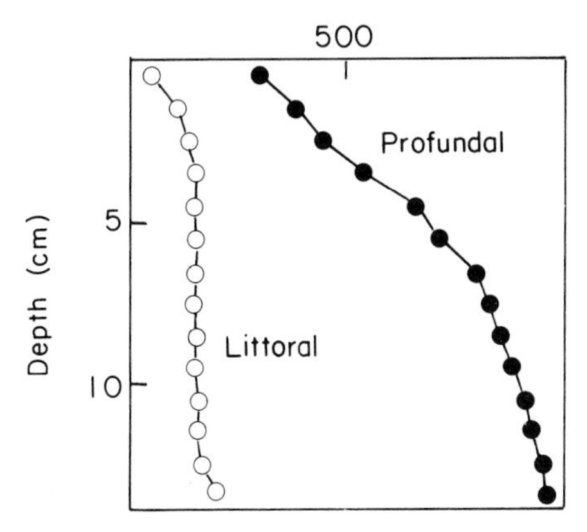

Fig. 13 Distribution of ammonium concentration with depth in the interstitial water of profundal (●) and littoral (○) sediments of a stratified eutrophic lake. Data from Jones and Simon [1981].

These are comparable to rates reviewed by Kamp-Neilson and Anderson [1977] for sediment water exchange and within the range 1.5 to 6.0 mmol m^{-2} day^{-1} obtained by Graetz *et al.* [1973] for Wisconsin lakes.

Using techniques similar to those described by Jørgensen [1978] for marine sediments, it is possible to demonstrate peaks of SO_4^{2-} reducing activity at the surface of profundal sediment in late summer (Fig. 14a). Activity in the littoral sediment is much less and a small sub-surface peak is observed. The interstitial concentrations of S^{2-} (Fig. 14b) are what might be expected from such a distribution of activity and of the population of sulphate reducing bacteria (Fig. 15). It is not possible to determine, from the relative rates of SO_4^{2-} disappearance and S^{2-} appearance, whether anaerobic decomposition of organic matter (putrefaction) plays a significant part in the production of S^{2-}. Molongoski and Klug [1980] believe that with the rapid removal of sulphate, putrefaction may be an important source of S^{2-} in the hypereutrophic Wintergreen Lake. In Blelham Tarn, field data

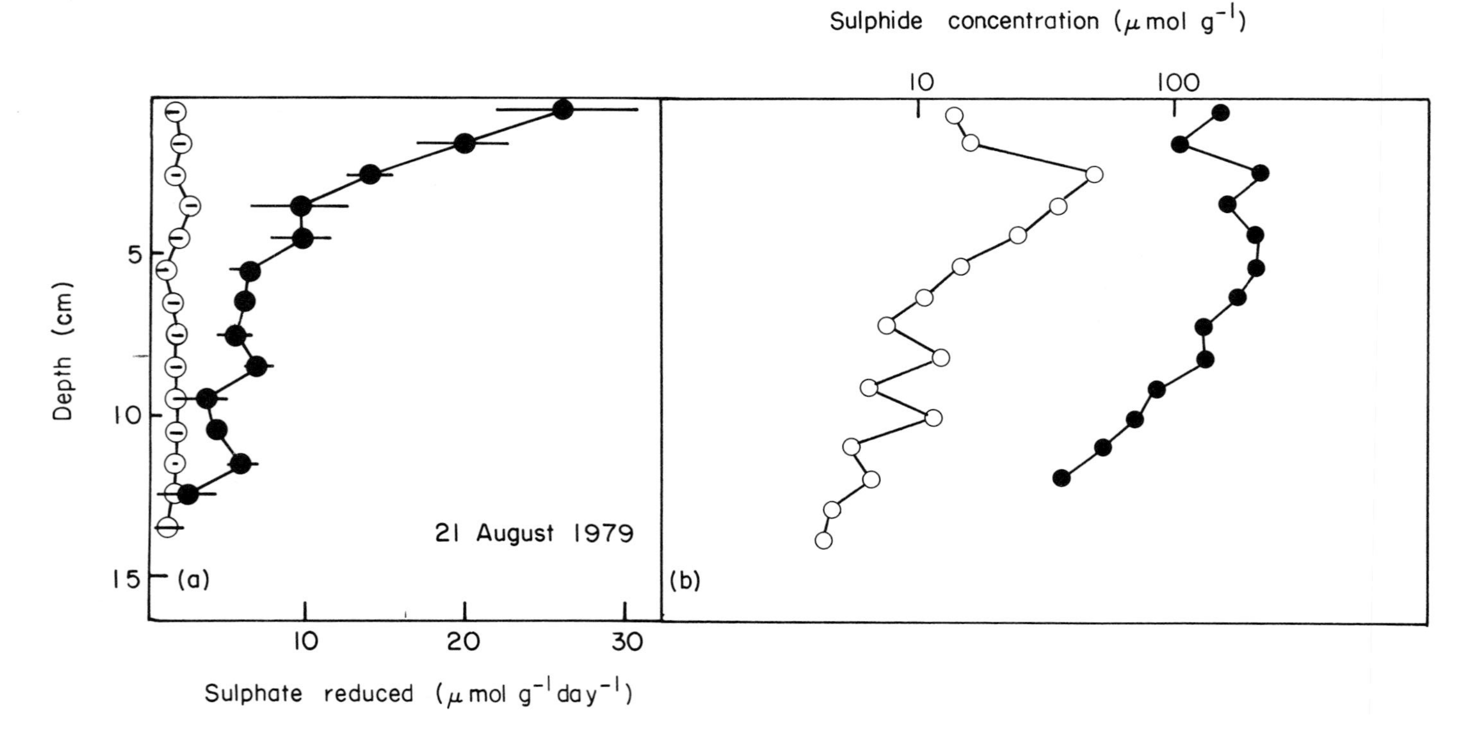

Fig. 14 **(a)** Rates of sulphate reduction in profundal (●) and littoral (○) sediments of a stratified eutrophic lake in late summer. Horizontal bars indicate 95% confidence limits. Those for the littoral samples fall within the width of the symbol. **(b)** Concentration of labile sulphide in the sediments. Data from Jones and Simon [1981].

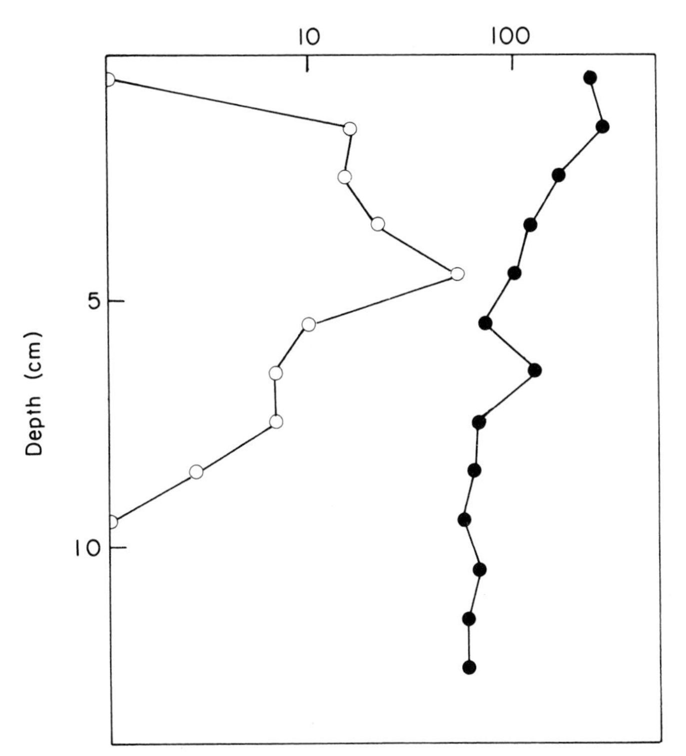

Fig. 15 Distribution of the MPN of sulphate reducing bacteria with depth in profundal (●) and littoral (○) sediments of a stratified eutrophic lake in late summer. Data from Jones and Simon [1981].

suggested that only 50% of the S^{2-} which appeared could be explained by SO_4^{2-} removal but the errors on such determinations are fairly large. Preliminary investigations have revealed relatively small numbers of putrefying bacteria and little capacity to release $^{35}S^{2-}$ from ^{35}S-methionine.

In the absence of O_2 and other electron acceptors, CH_4 becomes a major end product of decomposition in eutrophic lakes, accounting for more than half of the carbon input [Fallon *et al.*, 1980; Robertson, 1979; Rudd and Hamilton, 1978]. The methane concentration in the interstitial water of littoral and profundal sediments (Fig. 16a) is an indication of the relative rates of its production at these sites. These concentrations also correlate reasonably well with levels of Coenzyme F_{420} and are probably a good indication of the distribution of methanogens in the sediment (Fig. 16b). As summer progresses, the importance of methanogenesis in the surface sediment increases, and *in situ* release rates have been shown to correlate well with laboratory measurements of methanogenesis from slurries [Kelly and Chynoweth, 1980].

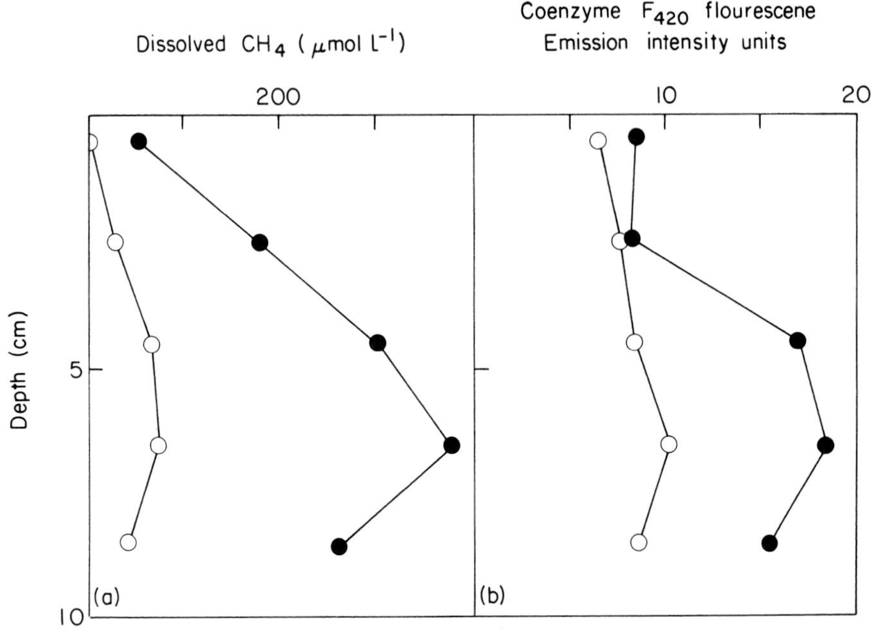

Fig. 16 **(a)** Concentrations of dissolved CH_4 and **(b)** relative fluorescence intensity of coenzyme F_{420} in profundal (●) and littoral (○) sediments of a stratified eutrophic lake.

Apart from experimental measurements of anaerobic processes in intact cores and with sediment slurries, it is also possible to make estimates of the activity of the benthic microbiota from the concentration gradients of key electron acceptors or products in the sediment. Space does not allow a detailed discussion of this approach but further information may be found on its application to inorganic nitrogen transformations in Vanderborght and Billen [1975], Bender *et al.* [1977] and Billen [this volume], to the kinetics of diagenesis nitrogen, sulphur, phosphorus and silicon in Berner [1974] and to methanogenesis in Bernard [1979]. The method has been applied largely to marine sediments, where gradients extend over considerable depths but it is worthy of consideration for the study of the freshwater environment.

Experimentally derived rates such as those described in this section, permit comparison of the relative importance of the various decomposition processes over the whole lake. Jones and Simon [1981] performed such an exercise on rates observed in the sediments of Blelham

TABLE 4

Comparison of decomposition processes in littoral and profundal sediments during summer stratification in a eutrophic lake

		Littoral*	Profundal+	Relative contribution to the whole lake ǂ Littoral : Profundal
CO_2 evolution		1600	940 - 1800+	1.2 : 1
§	O_2 uptake	1280	750 - 980+	1.2 : 1
§	NO_3^- reduction	320	220+ - 230	1.2 : 1
§	NH_4^+ accumulation	460	310+ - 320	1.2 : 1
§	N_2 evolution	115	30+ - 310	3.1 : 1
§	SO_4^{2-} reduction	15	30 - 55+	0.3 : 1
CH_4 evolution		50	80+ - 420	0.5 : 1

* With the exception of N_2 and CH_4 evolution, all rates for the littoral zone were derived from experimental systems.

\+ The values for the profundal zone indicate the ranges over three seasons. The individual values marked thus are those which are directly comparable with the values in the littoral zone.

ǂ Based on the areas of anoxic and oxygenated sediment at the end of the period of stratification.

§ Converted to CO_2 equivalent according to the stoichiometric equations of Richards [1965].

Data from Jones and Simon [1981].

Tarn; the results are summarized in Table 4. Although one might expect the profundal zone to be the major site of decomposition, it is clear from Table 4 that this is only true of SO_4^{2-} reduction and methanogenesis. Denitrification is three times more important, on a whole lake scale, in the littoral zone. The balance of processes in the two zones reflects, on the one hand, the development of sufficiently reducing conditions in the profundal sediments and, on the other, the increased temperature and replenishment of all electron acceptors in the littoral zone during the summer. The nature and availability of the carbon input at the two sites may also be important. These aspects of decomposition require further investigations of the kind reported by Klug *et al.* [1980].

Interactions between Benthic Bacteria

Just as the studies of Bryant and his coworkers [Bryant *et al.*, 1977; McInerney *et al.*, 1979; Boone and Bryant, 1980] have emphasized the importance of interspecies hydrogen transfer, the work of Cappenberg [1974*a,b*] and Cappenberg and Prins [1974] has probably done more than any other to stimulate interest in the experimental study of interactions between anaerobes, particularly methanogens and sulphate reducers, in freshwater sediments. Details of such interactions in marine systems will be dealt with elsewhere in this volume, and therefore this discussion will be confined to a brief review of the factors which are important in lakes.

In estuarine and marine systems the SO_4^{2-} concentration is such that SO_4^{2-} reducers and methanogens compete for H_2 with the former usually being the more successful [Abram and Nedwell, 1978]. The freshwater sediments very rapidly become depleted in SO_4^{2-} in summer and therefore the SO_4^{2-} reducing bacteria become net donors of H_2 to the methanogens. Thus the addition of SO_4^{2-} to such sediments inhibits methanogenesis [Winfrey and Zeikus, 1977] and this inhibition is reversed by the addition of H_2 or acetate [Mah *et al.*, 1977]. Cappenberg [1974*a*] concluded that methanogenesis was inhibited by excess S^{2-} accumulation, and that much of the methane produced was derived from acetate (an end-product of lactate metabolism by the sulphate reducers). With the wisdom of hindsight these conclusions might be questioned, but one can not deny the impetus which Cappenberg's work gave to research on anaerobic freshwater sediments [Winfrey *et al.*, 1977; Winfrey and Zeikus, 1979*a,b*; Strayer and Tiedje, 1978; Zaiss and Kaltwasser, 1979; Zeikus, 1977].

The actual precursors of methane in freshwater sediments have also received some attention in recent years. In his review, Zeikus [1977] draws on evidence from the work of Cappenberg [1974*b*] to suggest that acetate is the major precursor, but this and other earlier work was done with sediment slurries which had been gassed with N_2. Under such circumstances the sample is rapidly depleted in H_2 (as it is if sediment slurries are merely exposed to a gaseous phase) and therefore the $^{14}CH_2$ derived from added $^{14}CO_2$ will be underestimated. This suggests that sediment slurries are more appropriate for long term incubations in which the natural H_2 concentration may be reestablished, as, for example, in the procedures used by Abram and Nedwell [1978]. Accurate short term measurements are more likely to be obtained in core experiments such as those used by Jørgensen [1977] to determine rates of SO_4^{2-} reduction. More recent experiments have confirmed the importance of H_2 and CO_2 as CH_4 precursors [Strayer and Tiedje, 1978; Winfrey and Zeikus, 1979*b*; Zaiss, 1981] and experiments at this laboratory, using small cores and injecting them with traces of $^{14}CH_3COOH$ and

$^{14}CO_2$, suggest that two thirds of the CH_4 is derived from CO_2.

Interactions between aerobes and anaerobes also occur in sediments and this might be illustrated by returning to an examination of the fate of NO_3^- in littoral and profundal sediments. There is evidence that NH_4^+ rather than N_2 is a major end product of NO_3^- reduction at both sites, although experiments with ^{15}N are required to provide absolute proof of this. Ammonium has been shown to be an important end product in marine sediments [Sørensen, 1978*b*] and has received more attention recently [Cole and Brown, 1980; Caskey and Tiedje, 1980] particularly in relation to conditions of oxygen limitation [Dunn *et al.*, 1979]. Why, therefore, should N_2 gas release (denitrification) be significantly greater at the littoral site (Table 4)? Because the sediments are better aerated, and the Eh gradient is deeper, it would appear that higher and continued levels of nitrification in littoral sediments are responsible for recycling much of the NH_4^+ (Table 5). Even if relatively constant proportions of NO_3^- are converted to N_2 and NH_4^+ the recycling of the

TABLE 5

Relative importance of components of the nitrogen cycle in littoral and profundal sediments in Blelham Tarn

Approximate Rates mmol m^{-2} day^{-1}

Littoral	Profundal
Particulate nitrogen input << 3.0*	Particulate nitrogen input = 3.0

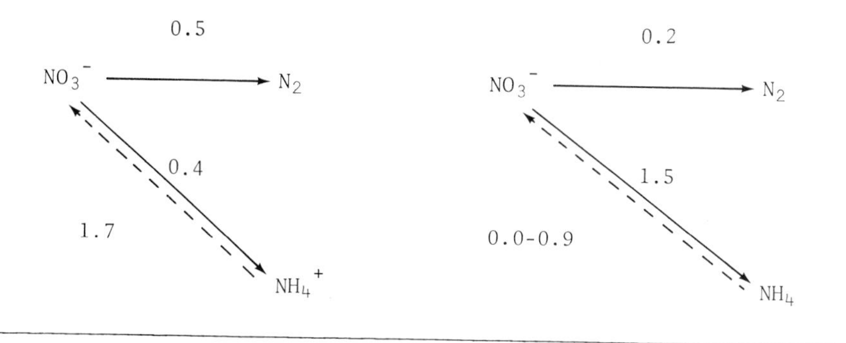

* Assumed to be less than that in the profunded zone because of resuspension.
Data from Jones and Simon [1981].

latter would automatically result in net enrichment in N_2. The role of nitrification in nitrogen cycling in lakes should not, therefore, be underestimated. Oxidation of NH_4^+ in the profundal zone is quantitatively less important at the beginning of the season, and stops as soon as the hypolimnion becomes deoxygenated. The coupling of these aerobic and anaerobic processes in sediment systems has been reported [Knowles, 1979] and the results in Table 4 are consistent with other reports of higher rates of denitrification in epilimnetic sediments [Chan and Campbell, 1980]. The stimulation of denitrification and nitrification by the activity of sediment macrofauna (bioturbation) has been reported [Chatarpaul *et al.*, 1980]. The numbers and variety of such animals are greater in the littoral zones of shallow eutrophic lakes [Jones, 1980] and may have contributed to these results.

Conclusions

The importance of aerobic and anaerobic bacterial metabolism in the major geochemical cycles in freshwater sediments has been demonstrated; it only remains to suggest research areas which might benefit from further investigation. These include:

(a) the role of Fe^{3+} reduction in organic matter decomposition.
(b) the relative importance of the various sediment zones in lakes, particularly those associated with the metalimnion.
(c) the importance of chemolithotrophic processes in recycling reduced products, particularly those of the sulphur cycle.
(d) the effect of microbial interaction such as bacterial grazing by protozoa on selected geochemical cycles.
(e) the importance of processes other than anaerobic respirations in the production of reduced N and S species.
(f) the importance of fermentative bacteria in C flow in anaerobic sediments.
(g) factors affecting the preservation of C and other elements in anoxic sediments.

These are the topics which spring to mind immediately, doubtless there are others, enough to keep many a freshwater microbiologist busy, and happy, for decades to come.

Acknowledgements

I am grateful to: B.M. Simon who has provided technical assistance of the highest quality in the study of some of the problems discussed in this chapter; D.W. Sutcliffe, S.I. Heaney and J.F. Talling who made available data on the chemistry of the lakes; the owners of the lakes for access to them; the Freshwater Biological Association for the opportunity and pleasure of working on them;

N.E.R.C. for financing this research; S.E. Whitehead for drawing the figures and J.C. Rhodes for typing the script.

References

Abram, J.W. and Nedwell, D.B. (1978). Inhibition of methanogenesis by sulphate reducing bacteria competing for transferred hydrogen. *Archives of Microbiology* **117**, 89-92.

Balashova, V.V. and Zavarzin, G.A. (1979). Anaerobic reduction of ferric iron by hydrogen bacteria. *Microbiology* **48**, 773-778.

Barber, L.E. and Ensign, J.C. (1979). Methane formation and release in a small Winsconsin lake. *Geomicrobiological Journal* **1**, 341-353.

Bender, M.L., Fanning, K.A., Froelich, P.N., Heath, G.R. and Maynard, V. (1977). Interstitial nitrate profiles and oxidation of sedimentary organic matter in the eastern equatorial Atlantic. *Science* **198**, 605-609.

Berner, R.A. (1974). Kinetic models for the early diagenesis of nitrogen, sulphur, phosphorus, and silicon in anoxic marine sediments. In "The Sea" (Ed. E.D. Goldberg), Vol. 5, pp.427-450. Wiley Interscience, New York.

Bernard, B.B. (1979). Methane in marine sediments. *Deep-Sea Research* **26A**, 429-443.

Boone, D.R. and Bryant, M.P. (1980). Propionate-degrading bacterium, *Syntrophobacter wolinii* sp. nov. gen. nov., from methanogenic ecosystems. *Applied and Environmental Microbiology* **40**, 626-632.

Burns, N.M. and Ross, C. (1972). "Project Hypo". Canada Centre for Inland Waters, Paper No.6. United States Environmental Protection Agency, Technical Report TS-05-71-208-24.

Bryant, M.P., Campbell, L.L., Reddy, C.A. and Crabill, M.R. (1977). Growth of *Desulfovibrio* in lactate or ethanol media low in sulphate in association with H_2-utilizing methanogenic bacteria. *Applied and Environmental Microbiology* **33**, 1162-1169.

Cappenberg, T.E. (1974*a*). Interrelations between sulphate-reducing and methane-producing bacteria in bottom deposits of a fresh-water lake. I. Field observations. *Antonie van Leeuwenhoek* **40**, 285-295.

Cappenberg, T.E. (1974*b*). Interrelations between sulphate-reducing and methane-producing bacteria in bottom deposits of a fresh-water lake. II. Inhibition experiments. *Antonie van Leeuwenhoek* **40**, 297-306.

Cappenberg, T.E. and Prins, R.A. (1974). Interrelations between sulphate-reducing and methane-producing bacteria in bottom deposits of a fresh-water lake. III. Experiments with ^{14}C-labelled substrates. *Antonie van Leeuwenhoek* **40**, 457-469.

Cappenberg, T.E. and Jongejan, E. (1978). Micro-environments for sulphate reduction and methane production in freshwater sediments. In "Environmental Biogeochemistry and Geomicrobiology" (Ed. W.E. Krumbein), Vol.1, The Aquatic Environment, pp.129-138. Ann Arbor Science Publishers, Michigan.

Caskey, W.H. and Tiedje, J.M. (1980). The reduction of nitrate to ammonium by a *Clostridium* sp. isolated from soil. *Journal of General Microbiology* **119**, 217-223.

Chan, Y.K. and Campbell, N.E.R. (1980). Denitrification in Lake 227 during summer stratification. *Canadian Journal of Fisheries and Aquatic Sciences* **37**, 506-512.

Charlton, M.N. (1980). Hypolimnion oxygen consumption in lakes: discussion of productivity and morphometry effects. *Canadian Journal of Fisheries and Aquatic Sciences* **37**, 1531-1539.

Chatarpaul, L., Robinson, J.B. and Kaushik, N.K. (1980). Effects of tubificid worms on denitrification and nitrification in stream sediment. *Canadian Journal of Fishery and Aquatic Sciences* **37**, 656-663.

Chen, R.L., Keeney, D.R., Konrad, J.G., Holding, A.J. and Graetz, D.A. (1972). Gas production in sediments of Lake Mendota, Wisconsin. *Journal of Environmental Quality* **1**, 155-157.

Clark, A.E. and Walsby, A.E. (1978*a*). The occurrence of gas-vacuolate bacteria in lakes. *Archives of Microbiology* **118**, 223-228.

Clark, A.E. and Walsby, A.E. (1978*b*). The development and vertical distribution of populations of gas-vacuolate bacteria in a eutrophic, monomictic lake. *Archives of Microbiology* **118**, 229-233.

Cole, J.A. and Brown, C.M. (1980). Nitrite reduction to ammonia by fermentative bacteria: a short circuit in the biological nitrogen cycle. *FEMS Microbiology Letters* **7**, 65-72.

Collins, V.G., Jones, J.G., Hendrie, M.S., Shewan, J.M., Wynn-Williams, D.D. and Rhodes, M.E. (1973). Sampling and estimation of bacterial populations in the aquatic environment. In "Sampling - Microbiological Monitoring of Environments" (Eds. R.G. Board and D.W. Lovelock) Society for Applied Bacteriology Technical Series, No. 7, pp.77-110. Academic Press, London.

Davison, W., Heaney, S.I., Talling, J.F. and Rigg, E. (1980). Seasonal transformations and movements of iron in a productive English lake with deep-water anoxia. *Schweizerische Zeitschrift für Hydrologie* **42**, 196-224.

Dick, M.W. (1971). The ecology of Saprolegniaceae in lentic and littoral muds with a general theory of fungi in the lake ecosystem. *Journal of General Microbiology* **65**, 325-337.

Dunn, G.M., Herbert, R.A. and Brown, C.M. (1979). Influence of oxygen tension on nitrate reduction by a *Klebsiella* sp. growing in chemostat culture. *Journal of General Microbiology* **112**, 379-383.

Edberg, N. (1976). Oxygen consumption of sediment and water in certain selected lakes. *Vatten* **32**, 2-12.

Edberg, N. and Hofsten, B.V. (1973). Oxygen uptake of bottom sediments studied *in situ* and in the laboratory. *Water Research* **7**, 1285-1294.

Edmondson, W.T. (1966). Changes in oxygen deficit of Lake Washington. *Verhandlungen der Internationalen Vereinigung für theoretische und angewandte Limnologie* **16**, 153-158.

Edwards, R.W. and Rolley, H.L.J. (1965). Oxygen consumption of river muds. *Journal of Ecology* **53**, 1-19.

Elliott, J.M. and Tullett, P.A. (1978). A bibliography of samplers for benthic invertebrates. Freshwater Biological Association, Occasional Publication No.4, p.61. Ambleside.

Fallon, R.D., Harrits, S., Hanson, R.S. and Brock, T.D. (1980). The role of methane in internal carbon cycling in Lake Mendota during summer stratification. *Limnology and Oceanography* **25**, 357-360.

Fenchel, T. (1972). Aspects of decomposer food chains in marine benthos. *Verhandlungsbericht der Deutschen Zoologischen Gesselschaft 65 Jahresversammlung* **14**, 14-22.

Fenchel, T.M. (1978). The microbial cycling of carbon, nitrogen and sulphur. *Society for General Microbiology Quarterly* **6**, 7-8.
Fenchel, T. and Blackburn, T.H. (1979). "Bacteria and Mineral Cycling" Academic Press, London.
Fenchel, T.M. and Jørgensen, B.B. (1977). Detritus food chains of aquatic ecosystems: the role of bacteria. *Advances in Microbial Ecology* **1**, 1-58.
Finlay, B.J. (1981). Oxygen availability and seasonal migrations of ciliated protozoa in a freshwater lake. *Journal of General Microbiology* (in press).
Gillespie, P.A. (1976). Heterotrophic potentials and trophic status of ten New Zealand lakes. *New Zealand Journal of Marine and Freshwater Research* **10**, 91-107.
Godinho-Orlandi, M.J.L. and Jones, J.G. (1981*a*). Filamentous bacteria in sediments of lakes of differing degrees of enrichment. *Journal of General Microbiology* **123**, 81-90.
Godhino-Orlandi, M.J.L. and Jones, J.G. (1981*b*). The distribution of some genera of filamentous bacteria in littoral and profundal lake sediments. *Journal of General Microbiology* **123**, 91-101.
Golterman, H.L. (1975). "Physiological Limnology". Elsevier, Amsterdam.
Golterman, H.L. (Ed.) (1977). "Interactions Between Sediments and Freshwater" Dr W. Junk, The Hague.
Graetz, D.A., Keeney, D.R. and Aspiras, R.B. (1973). Eh status of lake sediment-water systems in relation to nitrogen transformations. *Limnology and Oceanography* **18**, 908-917.
Hall, G.H., Collins, V.G., Jones, J.G. and Horsley, R.W. (1978). The effect of sewage effluent on Grasmere (English Lake District) with particular reference to inorganic nitrogen transformations. *Freshwater Biology* **8**, 165-175.
Hargrave, B.T. (1969). Epibenthic algal production and community respiration of Marion Lake. *Journal of the Fisheries Research Board of Canada* **26**, 2003-2026.
Hargrave, B.T. (1970). The effect of a deposit-feeding amphipod on the metabolism of benthic microflora. *Limnology and Oceanography* **15**, 21-30.
Hargrave, B.T. (1972). Aerobic decomposition of sediment and detritus as a function of particle surface area and organic content. *Limnology and Oceanography* **17**, 583-596.
Hargrave, B.T. (1975). The importance of total and mixed-layer depth in the supply of organic material to bottom communities. In "Limnology of Shallow Waters" (Ed. J. Satanski and J.E. Ponyi), pp.157-165. Akademia Kiado, Budapest.
Hargrave, B.T. (1976). The central role of invertebrate faeces in sediment decomposition. *Symposia of the British Ecological Society* **17**, 301-321.
Hayward, P. (1968). Hypolimnetic Oxygen Demand and Evolution of Gas from lake bottoms. Master's Report, University of North Carolina.
Heaney, S.I. and Talling, J.F. (1980). *Ceratium hirundinella* - ecology of a complex, mobile, and successful plant. *Report of the Freshwater Biological Association* **48**, 27-40.
Howard, D.L., Frea, J.I. and Pfiester, R.M. (1971). The potential for methane-carbon cycling in Lake Erie. *Proceedings of Conference on Great Lakes Research* **14**, 463-473.

Hutchinson, G.E. (1941). Limnological studies in Connecticut. IV. Mechanism of intermediary metabolism in stratified lakes. *Ecological Monographs* **11**, 21-60.

Hutchinson, G.E. (1957). "A Treatise on Limnology. Vol.I. Geography, Physics and Chemistry". Wiley, New York.

James, A. (1974). The measurement of benthal respiration. *Water Research* **8**, 955-959.

Johnston, D.W. and Cross, T. (1976). The occurrence and distribution of actinomycetes in lakes of the English Lake District. *Freshwater Biology* **6**, 457-463.

Johnston, D.W., Holding, A.J. and McCluskie, J.E. (1974). Preliminary comparative studies on denitrification and methane production in Loch Leven, Kinross and other freshwater lakes. *Proceedings of the Royal Society of Edinburgh* **B74**, 123-133.

Jones, J.G. (1976). The microbiology and decomposition of seston in open water and experiments enclosures in a productive lake. *Journal of Ecology* **64**, 241-278.

Jones, J.G. (1977*a*). The effect of environmental factors on estimated viable and total populations of planktonic bacteria in lakes and experimental enclosures. *Freshwater Biology* **7**, 67-91.

Jones, J.G. (1977*b*). The study of aquatic microbial communities. In "Aquatic Microbiology" (Eds. F.A. Skinner and J.M. Shewan), The Society for Applied Bacteriology Symposium Series No. 6, pp.1-30. Academic Press, London.

Jones, J.G. (1978). The distribution of some freshwater planktonic bacteria in two stratified eutrophic lakes. *Freshwater Biology* **8**, 127-140.

Jones, J.G. (1979*a*). Microbial activity in lake sediments with particular reference to electrode potential gradients. *Journal of General Microbiology* **115**, 19-26.

Jones, J.G. (1979*b*). Microbial nitrate reduction in freshwater sediments. *Journal of General Microbiology* **115**, 27-35.

Jones, J.G. (1980). Some differences in the microbiology of profundal and littoral lake sediments. *Journal of General Microbiology* **117**, 285-292.

Jones, J.G. (1981). The population ecology of iron bacteria (Genus Ochrobium) in a stratified eutrophic lake. *Journal of General Microbiology* **125**, 85-93.

Jones, J.G., Orlandi, M.J.L.G. and Simon, B.M. (1979). A microbiological study of sediments from the Cumbrian lakes. *Journal of General Microbiology* **115**, 37-48.

Jones, J.G. and Simon, B.M. (1980*a*). Decomposition processes in the profundal region of Blelham Tarn and the Lund Tubes. *Journal of Ecology* **68**, 493-512.

Jones, J.G. and Simon, B.M. (1980*b*). Variability in microbiological data from a stratified eutrophic lake. *Journal of Applied Bacteriology* **49**, 127-135.

Jones, J.G. and Simon, B.M. (1981). Differences in microbial decomposition processes in profundal and littoral lake sediments, with particular reference to the nitrogen cycle. *Journal of General Microbiology* **123**, 297-312.

Jørgensen, B.B. (1977). Distribution of colourless sulphur bacteria *(Beggiatoa* spp.) in a coastal marine sediment. *Marine Biology*, **41**, 19-28.

Jørgensen, B.B. (1978). A comparison of methods for the quantification of bacterial sulphate reduction in coastal marine sediments. I. Measurement with radiotracer techniques. *Geomicrobiology Journal* **1**, 11-27.

Jørgensen, B.B. (1980). Mineralization and the bacterial cycling of carbon, nitrogen and sulphur in marine sediments. In "Contemporary Microbial Ecology" (Eds. D.C. Ellwood, J.N. Hedger, M.J. Latham, J.M. Lynch and J.H. Slater), pp.239-251. Academic Press, London.

Kamp-Neilson, L. and Anderson, J.M. (1977). A review of the literature on sediment : water exchange of nitrogen compounds. *Progress in Water Technology* **8**, 393-418.

Kelly, C.A. and Chynoweth, D.P. (1980). Comparison of *in situ* and *in vitro* rates of methane release in freshwater sediments. *Applied and Environmental Microbiology* **40**, 287-293.

Kessel, J.F. (1978). The relation between redox potential and denitrification in a water-sediment system. *Water Research* **12**, 285-290.

Klug, M.G., King, G.M., Smith, R.L. and Lovley, D.R. (1980). Comparative aspects of anaerobic microbial metabolism in sediments along a transect of a lake basin. p. 94. *Abstracts of the Second International Symposium on Microbial Ecology, University of Warwick, UK.*

Knowles, R. (1979). Denitrification, acetylene reduction and methane metabolism in lake sediment exposed to acetylene. *Applied and Environmental Microbiology* **38**, 480-493.

Mah, R.A., Ward, D.M., Baresi, L. and Glen, T.L. (1977). Biogenesis of methane. *Annual Reviews of Microbiology* **31**, 309-343.

McInerney, M.J., Bryant, M.P. and Pfennig, N. (1979). Anaerobic bacterium that degrades fatty acids in syntrophic association with methanogens. *Archives of Microbiology* **122**, 129-135.

Molongoski, J.J. and Klug, M.J. (1980). Anaerobic metabolism of particulate organic matter in the sediments of a hypereutrophic lake. *Freshwater Biology* **10**, 507-518.

Mortimer, C.H. (1941). The exchange of dissolved substances between mud and water in lakes. I. Introduction. II. Changes in redox potential and in concentrations of dissolved substances in artificial mud-water systems, subjected to varying degrees of aeration. *Journal of Ecology* **29**, 280-329.

Mortimer, C.H. (1942). *Ibid.* III. The relation of seasonal variations in redox conditions in the mud to the distribution of dissolved substances in Esthwaite Water and Windermere North Basin. IV. General Discussion. *Journal of Ecology* **30**, 147-201.

Mortimer, C.H. (1971). Chemical exchanges between sediments and water in the Great Lakes - speculations on probable regulatory mechanisms. *Limnology and Oceanography* **16**, 387-404.

Munch, J.C. and Ottow, J.C.G. (1977). Model experiments on the mechanism of bacterial iron reduction in water-logged soils. *Zeitschrift für Pflanzenernahrung, Dungung and Bodenkunde* **140**, 549-562.

Odum, E.P. (1971). "Fundamentals of Ecology" W.B. Saunders Co., Philadelphia.

Ohle, W. (1956). Bioactivity, production and energy utilization of lakes. *Limnology and Oceanography* **1**, 139-149.

Oremland, R.S. (1979). Methanogenic activity in plankton samples and fish intestine: A mechanism for *in situ* methanogenesis in

oceanic surface waters. *Limnology and Oceanography* **24**, 1136-1141.
Ottow, J.C.G. and Glathe, H. (1971). Isolation and identification of iron-reducing bacteria from gley soils. *Soil Biology and Biochemistry* **3**, 43-55.
Ottow, J.C.G. and Munch, J.C. (1978). Mechanisms of reductive transformations in the anaerobic microenvironment of hydromorphic soils. In "Environmental Biogeochemistry and Geomicrobiology" (Ed. W.E. Krumbein) Vol. 2, pp.483-491. Ann Arbor Science Publishers, Michigan.
Powell, T. and Jassby, A. (1974). The estimation of vertical eddy diffusivities below the thermocline in lakes. *Water Resources Research* **10**, 191-198.
Revsbech, N.P., Jørgensen, B.B. and Blackburn, T.H. (1980). Oxygen in the sea bottom measured with a microelectrode. *Science* **207**, 1355-1356.
Richards, F.A. (1965). Anoxic basins and fjords. Chemical Oceanography **1**, (Eds. J.P. Riley and G. Skirrow), pp.611-695. Academic Press, London.
Robertson, C.K. (1979). Quantitative comparison of the significance of methane in the carbon cycles of two small lakes. *Ergebnisse der Limnologie* **12**, 123-135.
Rudd, J.W.M. and Hamilton, R.D. (1978). Methane cycling in a eutrophic shield lake and its effects on whole lake metabolism. *Limnology and Oceanography* **23**, 337-348.
Smith, I.R. (1975). Turbulence in Lakes and Rivers. Freshwater Biological Association Scientific Publication No.29.
Snodgrass, W.J. (1976). Potential effect of methane oxidation and nitrification-denitrification on the oxygen budget of Hamilton harbour. *Proceedings 11th Symposium, Water Pollution Research, Canada,* 101-107.
Sørensen, J. (1978*a*). Denitrification rates in marine sediment as measured by the acetylene inhibition technique. *Applied and Environmental Microbiology* **36**, 139-143.
Sørensen, J. (1978*b*). Capacity for denitrification and reduction of nitrate to ammonia in coastal marine sediment. *Applied and Environmental Microbiology* **35**, 301-305.
Spencer, M.J. (1978). Microbial activity and biomass relationships in 26 oligotrophic to mesotrophic lakes in South Island, New Zealand. *Verhandlungen der Internationalen Vereinigung für theoretische und angewandte Limnologie* **20**, 1175-1181.
Strayer, R.F. and Tiedje, J.M. (1978). *In situ* methane production in a small, hypereutrophic, hard-water lake: loss of methane from sediments by diffusion and ebullition. *Limnology and Oceanography* **23**, 1201-1206.
Takai, Y. and Kamura, T. (1966). The mechanism of reduction in waterlogged paddy soil. *Folia Microgiologica* **11**, 304-313.
Thauer, R.K., Jungermann, K. and Decker, K. (1977). Energy conservation in chemotrophic anaerobic bacteria. *Bacteriological Reviews* **41**, 100-180.
Trimble, R.B. and Ehrlich, H.L. (1970). Bacteriology of Manganese Nodules IV. Induction of an MnO_2-Reductive System in a Marine Bacillus. *Applied Microbiology* **19**, 966-72.
Vanderborght, J.-P. and Billen, G. (1975). Vertical distribution of nitrate concentration in interstitial water of marine sediments

with nitrification and denitrification. *Limnology and Oceanography* **20**, 953-961.

Verdouw, H. and Dekkers, E.M.J. (1980). Iron and manganese in Lake Vechten (The Netherlands); dynamics and role in the cycle of reducing power. *Archiv für Hydrobiologie* **89**, 509-532.

Vosjan, J.H. and Olanczuk-Neyman, K.M. (1977). Vertical distribution of mineralization processes in tidal sediment. *Netherlands Journal of Sea Research* **11**, 14-23.

Wetzel, R.G., Rich, P.R., Miller, M.C. and Allen, H.L. (1972). Metabolism of dissolved and particulate detrital carbon in a temperate hard-water lake. *Memorie dell Instituto Italiano di Idrobiologia Dott. Marco de Marchi* **29**, (Supplement) 185-243.

Wieser, W. and Zech, M. (1976). Dehydrogenases as tools in the study of marine sediments. *Marine Biology* **36**, 113-122.

Willoughby, L.G. (1961). The ecology of some lower fungi at Esthwaite Water. *Transactions of the British Mycological Society* **44**, 305-332.

Willoughby, L.G. (1965). Some observations on the location of fungal activity at Blelham Tarn. *Hydrobiologia* **25**, 352-356.

Willoughby, L.G. (1969). A study of the aquatic actinomycetes of Blelham Tarn. *Hydrobiologia* **34**, 465-483.

Willoughby, L.G. (1974). The ecology of lower freshwater phycomycetes in the tube experiment at Blelham Tarn. *Veroffentlichungen des Instituts für Meeresforschung in Bremerhaven, Supplement* **5**, 175-195.

Winfrey, M.R., Nelson, D.R., Klevickis, S.C. and Zeikus, J.G. (1977). Association of hydrogen metabolism with methanogenesis in Lake Mendota sediments. *Applied and Environmental Microbiology* **33**, 312-318.

Winfrey, M.R. and Zeikus, J.G. (1977). Effects of sulphate on carbon and electron flow during microbial methanogenesis in freshwater sediments. *Applied and Environmental Microbiology* **33**, 275-281.

Winfrey, M.R. and Zeikus, J.G. (1979*a*). Microbial methanogenesis and acetate metabolism in a meromictic lake. *Applied and Environmental Microbiology* **37**, 213-221.

Winfrey, M.R. and Zeikus, J.G. (1979*b*). Anaerobic metabolism of immediate methane precursors in Lake Mendota. *Applied and Environmental Microbiology* **37**, 244-253.

Woolfolk, C.A. and Whiteley, H.R. (1962). Reduction of inorganic compounds with molecular hydrogen by *Micrococcus lactilyticus*. *Journal of Bacteriology* **84**, 647-658.

Zaiss, U. (1981). Seasonal studies of methanogenesis and desulfurication in sediments of the River Saar. Zentralblatt für Bakteriologie, Parasitenkunde, Infektionskrankheiten und Hygiene (Abteilung I) (in press).

Zaiss, U. and Kaltwasser, H. (1979). Hydrogenase activity and methanogenesis in anaerobic sewage sludge, in rumen liquid, and in freshwater sediments. *European Journal of Applied Microbiology and Biotechnology* **8**, 217-227.

Zeikus, J.G. (1977). The biology of methanogenic bacteria. *Bacteriological Reviews* **41**, 514-541.

Chapter 6

MICROBIAL ACTIVITY IN ORGANICALLY ENRICHED MARINE SEDIMENTS

N.S. BATTERSBY and C.M. BROWN

Department of Brewing and Biological Sciences, Heriot-Watt University, Chambers Street, Edinburgh, Scotland, UK

Introduction

Microbial activity in marine sediments is highly variable being a function of sediment location (in shore, deep sea etc.), the rate of sedimentation, the depth of the water column, the chemical nature of the sedimenting material and the local hydrographic conditions. Some of the sea lochs (fjords) on the west coast of Scotland provide ideal sites for the study of processes in these sediments due to their ready accessibility. Loch Eil in Argyll is one such site and has been the subject of a large research effort for the last 15 years. The principal reason for this was the siting of a paper and pulp mill at the entrance of the loch and the subsequent discharge of waste cellulose fibre which, being heavier than sea water, rapidly settled out from the water column. The chemical and biological effects produced upon the microbial degradation of this cellulose fibre have been the subject of a number of publications, many of which are summarized or referred to in the paper of Pearson [1982].

This paper describes the microbiological degradation of the cellulose fibre, the fate of the products of this cellulolysis and compares data from Loch Eil with those obtained from other marine sediments.

The Loch Eil Ecosystem

The geographical location of Loch Eil with the positions of the main sampling stations and the paper and pulp mill are illustrated in Fig. 1. The loch is approximately 10 km long and 1 km wide and is separated from Loch Linnhe by the Annat narrows. The narrows form a sill between the basins of Loch Linnhe and Loch Eil which is important hydrographically as a barrier to the exchange of bottom water and as a region of localized mixing of

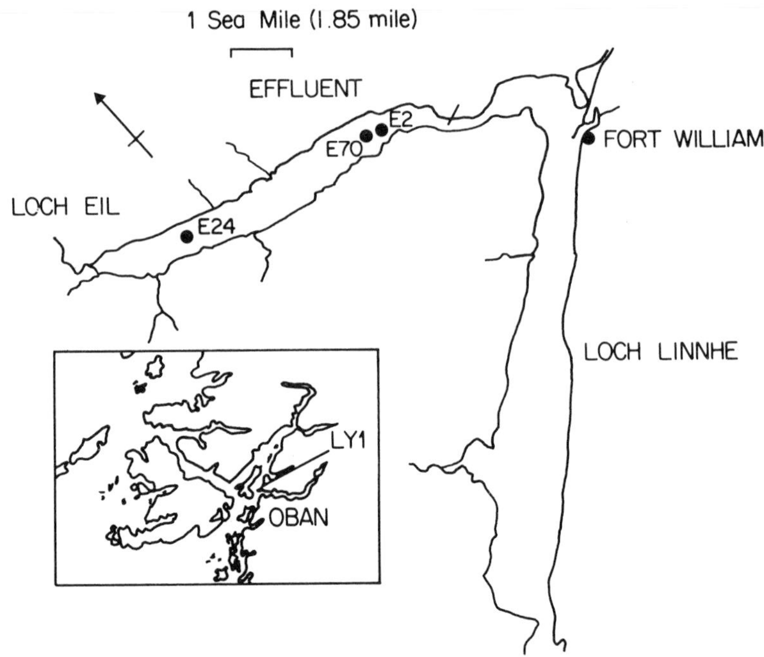

Fig. 1 Position of main sampling stations in Loch Eil and the Firth of Lorne.

surface and near surface waters. Station E24 is about 1.5 km from the head of the loch with a water column depth of approximately 32 m. The sediments of this station are less reduced than those at E70 and E2 and less affected by the cellulose effluent. Station E2 and E70 are in the deep basin where the water depth is approximately 50 m and are located closer to the effluent outfall. These sediments are highly reduced and rich in sulphide although the overlying water is well oxygenated. As a control station in the same vicinity but outside the influence of the pulp mill discharge, station LY1 is situated in the Lynn of Lorne with a water depth of 46 m. This sediment is well oxidized at the surface and is not affected by any cellulose-rich effluent [see Stanley *et al*., 1978, 1980]. Pearson [1982] has compiled figures for the total input of organic carbon to the loch system over a period of a year. His calculations indicate that terrestrial carbon input accounted for 4% (2 to 7% variation), planktonic carbon for 12% (1 to 28%) and effluent carbon input 84% (67 to 97%) total. This data emphasizes the overwhelming importance of the mill effluent on the loch system. Chemical analysis of the sediments indicated that the carbon content of the top

TABLE 1

Chemical composition of Loch Eil and N.E. Atlantic sediments

Station	Water depth (m)	Eh (mV)*	% Carbon	% Nitrogen	C:N	Cellulose (mg g^{-1} dry wt)
LY1	47	+29	2.9	0.37	7.84	0.62
E24	30	-56	5.3	0.46	11.46	3.29
E70	49	-158	6.2	0.46	13.43	4.33
A1	4,920	+525	0.3	-	-	nd
A2	2,880	+341	0.3	-	-	nd
A3	158	+365	0.5	-	-	2.6

* Eh measured at a depth of 4 cm

nd not detectable

3 cm of core samples was approximately 2 times higher in E24 and E70 than in equivalent samples from LY1. Similarly the cellulose contents of the Loch Eil sediments were some 5 to 6 times higher than in LY1. These data are summarized in Table 1. Cellulose fibre has been discharged into the Loch since 1966 and the content of cellulose in the sediments has been monitored continuously since 1972. While the discharge rates from the mill have shown wide fluctuations in this time period, there is no evidence of any appreciably long term increase in the cellulose content of the sediments. The natural system has been simulated in experimental tanks containing sediment and loaded daily with pulp fibre at a rate to match the observed rate of deposition into the stations in the deep basin of the loch. Data obtained over a six month period showed an initial increase in detectable levels of cellulose during what was apparently an adaptation period but a steady state situation was then reached in which the rate of cellulolysis equalled the rate of cellulose entering the system [Stanley *et al.*, 1978]. This indicated that simple chemical analysis of the sediments for cellulose gave no clear indication of the rate at which cellulose degradation was proceeding. The redox potential in the sediments of the experimental tanks following the start of daily loading with cellulose fibre decreased progressively to around -150 mV in the tank receiving cellulose fibre while in the control tank which was not loaded (Fig. 2) no such large decrease was observed. It

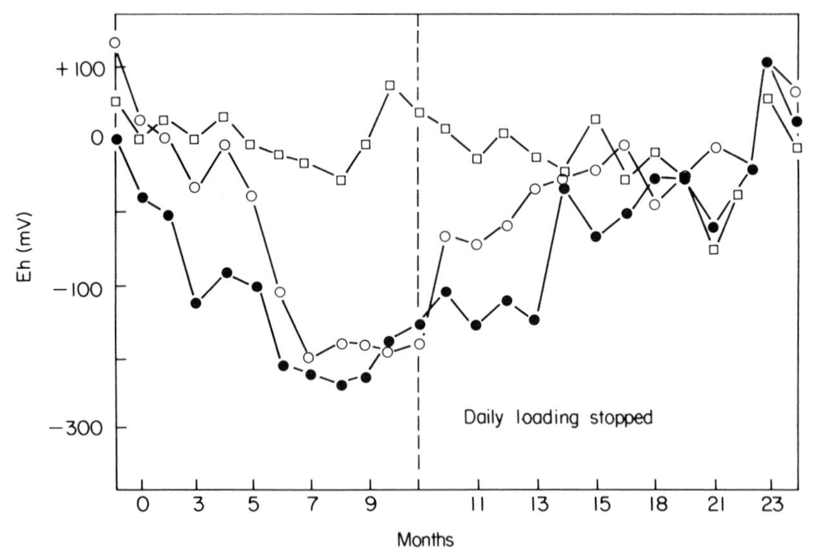

Fig. 2 Long-term changes in redox at 4 cm depths in experimental tanks loaded daily with cellulose pulp fibre compared with unloaded control tank. (Redox readings are uncorrected, add +198 mV to give a true reading.) ○ 0.89 g cellulose m^{-2} d^{-1}, ● 1.88 g cellulose m^{-2} d^{-1}, □ unloaded control tank.

was also shown that cessation of loading resulted in a progressive increase in redox potential back to the control level. In the loch itself zones of highly reduced sediments exist within the deep basin, the area of these zones being closely related to the rate of cellulose fibre discharge. The highly negative redox potential of these sediments was associated with a high concentration of sulphide indicating the presence of active populations of sulphate reducing bacteria [Pearson and Stanley, 1979; Stanley *et al.*, 1980].

Biological Status of Sediments

A survey of the bacterial populations in the LY1 and Loch Eil sediments indicated that no increase in the total numbers of aerobic and anaerobic heterotrophic, nor nitrate reducing, organisms occurred as a result of organic enrichment. In contrast, the numbers of sulphate reducing bacteria were on average 2 to 3 times, and those of cellulolytic bacteria up to 5 times, higher in E2 and E70 than in LY1 sediments (Table 2). In addition, Pearson [1982] has gathered together information on the numbers of ciliates, meiofauna and macrofauna in the sediments. These data indicate a considerable enrichment in the numbers of all three groups of animals present in those sediments which are highly enriched with cellulose fibre. For example, the size of the ciliate population, while being at a similar level in Stations LY1 and E24, was increased some 10 fold in Stations E2 and E70. Since the major food source of this fauna may be considered to be bacteria it follows that the rate of turnover (growth and grazing) of bacteria in these organically enriched sediments must be proportionally higher than those of Stations E24 and LY1 while the total bacterial numbers remained similar. Excessive loading of cellulose fibre into experimental tanks resulted in the formation of a fibrous mat at the sediment surface. When examined with the scanning electron microscope this was shown to be made up of fibres of cellulose interwoven with filaments of the sulphide oxidizing bacterium *Beggiatoa*. Observations with an underwater television camera showed extensive areas of mat formation over a large part of the deep basin of Loch Eil following a period of unusually high input of pulp fibre. *Beggiatoa* mats have been reported in other marine sediments subjected to large inputs of organic material [see Stanley *et al.*, 1978].

Cellulose Degradation in Marine Sediments

The bacterial degradation of cellulose has been studied both in the laboratory experimental tank systems and in sediment samples from Loch Eil and Station LY1. The bleached wood pulp used in this study was that produced by the Stora sulphite process and was supplied by

TABLE 2

Bacterial numbers present in Loch Eil sediments (g^{-1} dry sediment)

	Numbers in each station							
	E2		E70		E27		LY1	
	Mean	Range	Mean	Range	Mean	Range	Mean	Range
Cellulolytic bacteria	$3x10^3$	$3x10^3$ - $2.3x10^4$	$2.5x10^3$	$5x10^2$ - $1.8x10^4$	$7x10^2$	5 - $1.3x10^4$	70	3 - $6x10^2$
Sulphate reducing bacteria	$3.2x10^5$	$2x10^4$ - $6.2x10^5$	$2.9x10^5$	$4x10^4$ - $2.1x10^6$	$5x10^4$	$1x10^4$ - $1x10^5$	$7x10^4$	$1x10^4$ - $9x10^4$
Anaerobic heterotrophic bacteria	$3.6x10^5$	$1.2x10^5$ - $9.8x10^5$	$1.1x10^6$	$1.2x10^5$ - $6.3x10^6$	$1.5x10^5$	$3x10^4$ - $4.7x10^5$	$3.4x10^5$	$3x10^4$ - $8.1x10^5$

Scottish Pulp and Paper Mills Ltd., Fort William. Analysis of the pulp fibre showed it to be 98% cellulose, being 43% carbon and 0.7% nitrogen [Vance *et al.*, 1979]. The appearance of this pulp, as revealed by scanning electron microscopy, was that of individual wood cells in the size range 15 to 55 μm in width and up to 5 μm long. Pulp recovered from the sediment model showed a layer of bacteria on the outer surface of the wood cells and these bacteria were often associated with mucilage. Examination in the light microscope suggested that the bacteria gained access to the central lumena of the wood cells through both pit apertures and the broken ends of the cells. Transmission electron microscopy indicated that bacteria were associated with cavities in the exposed secondary wall of the pulp cells and in most cases the cell wall of the bacteria was surrounded by a network of fine projections which were interpreted as being extracellular polymeric materials. In most instances the bacteria were closely associated with obvious zones of cellulolysis but were not in direct contact with the substrate. The invasion of the wood pulp by the bacteria extended several micrometres into the secondary cell walls and the zones of cellulolysis apparently corresponded to the passage of individual bacteria. Extensive areas of cellulolysis were also observed in some regions of the pulp cell walls and were the results of the activity of several bacteria [Vance *et al.*, 1979]. The overall susceptibility of this wood pulp to bacterial degradation is thought to be due to the removal of lignin and an increase in the size of the cell wall capilliaries during the pulping process. The cellulose component of the wood may also undergo a reduction in the degree of polymerization during the pulping process and this increases the susceptibility of the material to the cellulolytic enzymes. At no stage in this investigation was there any indication of extensive cellulolysis due to fungi [Vance, 1977]. The numbers of aerobic and anaerobic cellulolytic bacteria were enumerated using the most probable number (MPN) method using chromatography paper as the source of cellulose. Cellulolytic activity was determined in sediment cores by inserting strips of chromatography paper attached to glass tubes, and in cultures of isolated bacteria by determining the rate of appearance of the dye remazol brilliant blue which had been used previously to stain the bleached wood pulp [Vance *et al.*, 1982].

The total population densities of cellulolytic bacteria observed in the present work are of the same order as those found earlier by Zobell [1938] and Goman [1973] when corrections are made for the water contents of the sediments. Cellulolysis occurred both aerobically and anaerobically. In the aerobic MPN tubes degradation of the chromatography paper was restricted to the medium interface while in the anaerobic tubes the whole area of paper submerged in the liquid showed discrete zones of cellulolysis. The preheating of sediments resulted in total loss of cellulolytic activity indicating that the active population in the culture tubes were not present as spores in the sediment. Data on the vertical distribution of cellulolytic bacteria indicated that maximum numbers of both aerobic and anaerobic organisms occurred near the sediment surface. While these organisms could also be recovered from sediment depths down to 5 cm, experiments involving the insertion of test strips of chromatography paper indicated that cellulolytic activity was restricted to the upper 1 cm of a sediment core. As was indicated earlier, the numbers of both aerobic and anaerobic cellulolytic bacteria were higher in sediments from Loch Eil than those at Station LY1 and within the loch the highest numbers of organisms were found persistently in the deep basin stations at E2 and E70. In most samples the aerobic bacteria were numerically dominant although the relative activities of the aerobic and anaerobic populations were not determined. There was however a good correlation at Station E2 between the number of aerobic cellulolytic bacteria and the mean effluent discharge during the previous month [Vance, 1977; Vance *et al.*, 1981]. Wood pulp of the type discharged by the mill was utilized as the role source of carbon and energy by mixed cultures of aerobic bacteria enriched from positive MPN tubes. It was interesting to note that pulp which had been labelled with remazol brilliant blue was more resistant to bacterial cellulolysis than the native fibre. Similarly the degradation rate of dried pulp fibre was much slower than that of the non-dried material, indicating that the physical characteristics of the substrate were important factors in determining the rates of degradation. As expected from the electron microscopy evidence, extracellular β1-4-glucanase activity was found in cultures of bacteria enriched from sediments of Station E70 with cellulose as carbon source, but no activity could be detected in unconcentrated samples of interstitial water from the sediment cores. While enzyme activity was detected after sample concentration it seems likely that *in situ* the enzymes were closely associated with the outer envelope layers of the cellulolytic bacteria.

The Products of Cellulose Degradation

The breakdown of cellulose in these sediments is assumed to occur by depolymerization reactions to produce cellibiose and glucose, both of which decreased the rate of cellulolysis when added to enrichment cultures of organisms obtained from the loch sediments [Vance, 1977]. The routes of further degradation of these mono and disaccharides to low molecular weight intermediates are indicated by the appearance of a range of organic acids (including acetate, propionate, butyrate, valerate, succinate and lactate) in the pore waters. The assays of these intermediates involved either squeezing core sediment samples with a core squeezer (or alternatively with a dialysis method) followed by gas chromatography or enzymic analysis [Miller *et al.*, 1979]. Acetate was the most prominent fatty acid in the surface layers of all those sediments studied. In addition, surface sediments from Station LY1 contained appreciable concentrations of lactate with trace amounts of butyrate and valerate. Analysis of the Loch Eil sediments, however, showed a 2 to 5 fold increase in the acetate concentration and also the appearance of high concentrations of propionate, butyrate, valerate, lactate and in Station E70 succinate within the surface layers. Further experiments with the experimental tank investigated the effects of the addition of different loadings of cellulose fibre on fatty acids production. An increase in the daily input of cellulose fibre of up to 2.2 g per square metre per day resulted in a steady increase in the acetate levels within the pore water (Table 3). The concentrations of the other organic acids were more variable but propionate, butyrate, lactate and succinate all accumulated at the higher cellulose concentrations. Miller *et al.* [1979] have discussed the possible origins of these organic acids, the concentrations of which reflect the rates of production and utilization of these compounds. Since the aerobic oxidation of these intermediates is unlikely, due to the low redox potential prevailing in the sediments, they presumably serve as electron donors for nitrate reduction and sulphate reduction. No significant production of methane is usually detected within marine sediments [see Fenchel and Jørgensen, 1977] due to sulphate reducing bacteria competing successfully for the available electron donors [acetate and hydrogen, Abram and Nedwell, 1978]. However under conditions of sulphate depletion [Martens and Berner, 1974] or in freshwater sediments [Jones and Simon, 1980, 1981] methanogenesis is of more importance and can account for up to 25% of carbon mineralized [Jones and Simon, 1981]. Within the sediments of Loch Eil methanogenesis did not play a major role in anaerobic oxidation of organic material.

In Loch Eil, as in many marine sediments, degradation of organic detritus by heterotrophic microorganisms results in the formation of an anoxic zone beneath the sediment surface. The depth of this zone depends in the main on the oxygen regime of the overlying water and the biodegradability of the organic material entering the sediment [Stanley *et al.*, 1978]. Such reduced areas can be mapped and their fluctuations followed by means of redox potential measurements [Pearson and Stanley, 1980]. In Loch Eil which receives a high rate of sedimenting, easily biodegradable organic material, typical Eh values of -150 mV at 4 cm depth were recorded. In contrast, sediments from the abyssal plain of the N.E. Atlantic (which has low rates of sedimentation of highly refractory

TABLE 3

Short-chain fatty acids from sediment pore water in experimental tanks loaded with different amounts of cellulose fibre

Daily loading	Fatty acid concentration ($\mu g\ ml^{-1}$ pore water)					
($g\ m^{-2}\ day^{-1}$)	Acetate	Propionate	Butyrate	Valerate	Lactate	Succinate
0	148.0	-	3.0	-	-	-
0.5	283.7	tr	10.0	29.0	tr	-
1.13	410	-	2.7	-	nd	nd
1.50	516.7	tr	0.2	-	nd	nd
2.25	244.0	3.7	8.0	tr	3.8	10.0

tr trace amounts found

\- not detected

nd not determined

organic material) had a redox potential of +525 mV at 4 cm depth [Stanley *et al.*, 1978]. In nearshore marine sediments the reduced zone occurs usually a few centimetres below the sediment surface.

Nitrate Reduction

The use of nitrate as terminal electron acceptor, and its subsequent reduction to nitrite, is a widespread property of bacteria and occurs in some 40 genera of facultatively anaerobic organisms and a few strict anaerobes [Payne, 1973]. This process, which is energetically superior to fermentative metabolism and to the use of sulphate or CO_2 as terminal electron acceptor, involves a membrane bound nitrate reductase whose synthesis is repressed by the presence of oxygen [Haddock and Jones, 1977]. The reaction follows the equation:

$$NO_3^- + H_2 \longrightarrow NO_2^- + H_2O \qquad \Delta G^{o1} = -34 \text{ kcal/mol}$$

Reported electron donors for nitrate reduction are hydrogen, formate, NADH, succinate, lactate and glycerol phosphate [Thauer *et al.*, 1977]. In several pseudomonads linear hydrocarbons and aromatic compounds can act as electron donors [Taylor and Heeb, 1972; Traxler and Bernard, 1969] with ring cleavage probably occurring via a benzoate intermediate [Taylor *et al.*, 1970]. The chemolithotroph *Thiobacillus denitrificans* utilizes sulphide, sulphite, thiosulphate and hydrosulphite as electron donors [Adams *et al.*, 1971].

A smaller variety of bacteria can reduce nitrite further, either to nitrogen or to ammonia. The reduction of nitrite to gaseous products (denitrification) involves nitrite, nitric oxide and nitrous oxide as electron acceptors in energy yielding reactions, with the gaseous products formed being lost to the atmosphere. This has been the conventional view of the reduction of nitrate in oxygen depleted water, sediments and soil [Cole and Brown, 1980] and rates of reduction $NO_3^- \rightarrow N_2$ of 0.87 µmol N ml^{-1} d^{-1} have been reported by Søorensen [1978*a*] for the top 3 cm of an inshore marine sediment. Several fermentative bacteria may also use nitrite as an oxidizing agent and reduce it rapidly to ammonia [Cole and Brown, 1980]. Evidence for the importance of ammonia formed from nitrate reduction in aquatic systems has been accumulating recently [see Herbert, this volume]. In the same report mentioned above Søorensen [1978*a*] measured rates of conversion of $NO_3^- \longrightarrow NH_4^+$ of the order of 0.75 µmol of N ml^{-1} d^{-1}, and in ^{15}N- tracer incubations of anaerobic coastal sediments Koike and Hattori [1978] showed that between 20% and 67% of the ^{15}N nitrate was converted to ammonia or particulate organic nitrogen.

Samples from the top 5 cm of Loch Eil sediment have been incubated with or without nitrate supplementation at *in situ* temperature for 1, 5 or 7 days. On supplementation with nitrate all stations produced nitrite after 24 hrs incubation with the highest activity being measured at Station E24 (0.025 μmol of N ml^{-1} d^{-1} $NO_3^- \rightarrow NO_2^-$ with 10 mM NO_3^- supplementation). The ammonia concentrations recorded were much more variable and in many instances decreased on incubation, presumably due to assimilation by sediment organisms. E70 sediment however did show net ammonia production which was in part stimulated by nitrate addition. Similar experiments on samples from LY1 and the North East Atlantic (A3 158 m, A2 2880 m and A1 4920 m) have shown that all stations produced nitrite after 24 hrs incubation on supplementation with nitrate, with the highest activity (0.026 μmol of N ml^{-1} d^{-1} $NO_3^- \rightarrow NO_2^-$ with 10 mM NO_3^- supplementation) at the continental shelf sampling station A3 and lowest activities in sediments from the abyssal Atlantic (0.005 μmol of N ml^{-1} d^{-1} at A2 and 0.001 μmol of N ml^{-1} d^{-1} $NO_3^- \rightarrow NO_2^-$ at A1, both with 10 mM NO_3^- supplementation. Again ammonia concentrations were variable, although net production (in part nitrate stimulated) occurred in A3 samples.

The numbers and types of nitrate-reducing flora in the above sediments have been investigated. Serial dilutions of the various sediments were plated out onto nitrate agar and incubated anaerobically for 7 days. After incubation and counting, 50 colonies were picked at random from each sediment, classified to genus level and their action on nitrate investigated. From the results in Table 4 it can be seen that highest numbers of nitrate reducing bacteria were recovered from the inshore (LY1, E24 and E70) and continental shelf (A3) stations, with fewer recovered from A2. The counts of nitrate reducing bacteria from the inshore marine sediments were much lower than those obtained from freshwaters (values of between 10^9 to 10^{10} nitrate reducers g^{-1} being observed by Jones and Simon [1981]) and may reflect the relative unimportance of nitrate reduction in marine sediments due to the low nitrate concentrations prevailing [Sørensen *et al.*, 1979]. In all stations except E70 the majority of organisms isolated were *Aeromonas* or *Vibrio* sp. and this preponderance of fermentative bacteria suggested a similar nitrate reducing flora to that active in estuarine sediments [Dunn *et al.*, 1980; Herbert *et al.*, 1980; Cole and Brown, 1980]. Ninety-two to 100% of the bacteria recovered by the above methods were able to reduce nitrate and 80 to 100% to reduce nitrite to ammonia. Reduction of nitrite to gaseous products was not encountered. Similar results have been reported by Nedwell [1975] in a study of the action on nitrate of heterotrophic bacteria isolated from a tropical mangrove estuary. Forty-three percent of the bacteria isolated reduced nitrate to nitrite, 30% reduced nitrite to ammonia and only 2% could denitrify nitrate to gaseous products. These data suggest that in many marine and estuarine environments nitrate reduction

TABLE 4

Bacterial counts, groups and action on nitrate of nitrate reducing bacteria from Loch Eil and N.E. Atlantic sediments

Station	Nitrate-reducing bacteria (ml^{-1})	Bacterial identification (% total)				Action on nitrate (% total)			
		A/V	PLO	Bacillus	Others	None	$NO_3^- \rightarrow NO_2^-$	$NO_3^- \rightarrow N_2$	$NO_3^- \rightarrow NH_4^+$
LY1	5×10^3	76	22	2	-	2	-	-	98
E24	5×10^3	88	12	-	-	2	-	-	98
E70	3×10^3	32	40	12	16	8	4	-	88
A2	1×10^2	100	-	-	-	-	-	-	100
A3	1.5×10^3	95	5	-	-	-	-	-	100

A/V *Aeromonas/Vibrio*

PLO pseudomonads

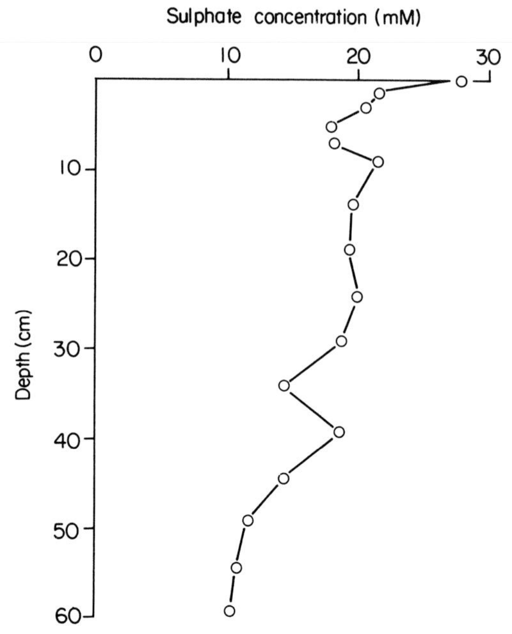

Fig. 3 Profile of sulphate concentration with depth at Station E70 Loch Eil Nov. 1979. [F. Drake, unpublished data]

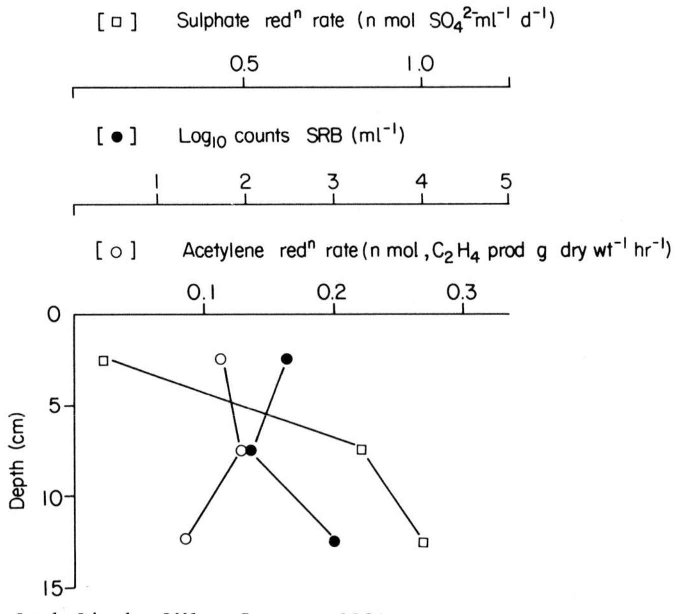

Fig. 4 Loch Linnhe LY1 - January 1981

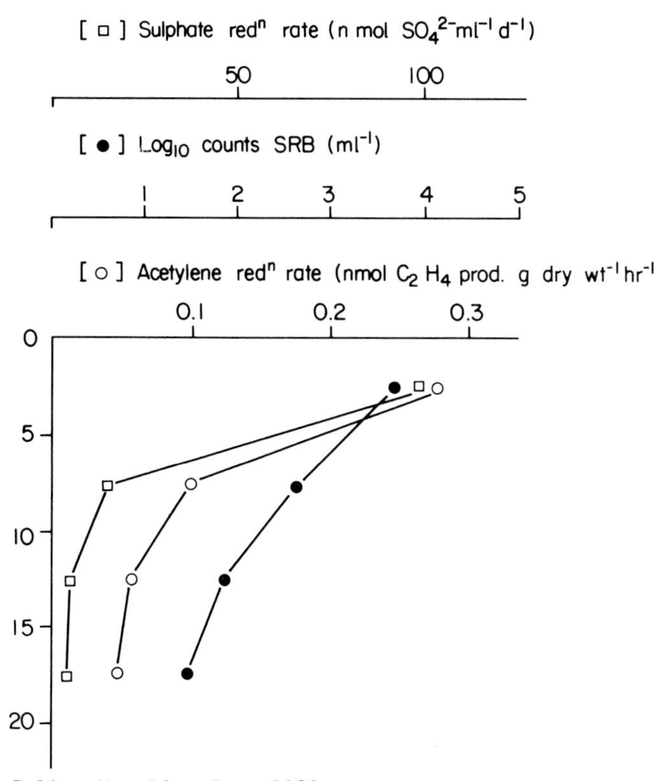

Fig. 5 Sullom Voe D4 - June 1980

results in nitrogen being conserved within the ecosystem as ammonia [see also Herbert, and Jones, this volume].

As mentioned elsewhere [Nedwell, this volume] nitrate reduction tends to occur between the aerobic zone and the zone of sulphate reduction in marine sediments. In most sediments this will be the upper 5 cm [for example Sørensen, 1978*a*,*b*] with nitrate reducing activity decreasing with depth down the sediment (due to nitrate depletion). In some sediments, nitrate reduction may be inhibited by sulphate reduction either by sulphide toxicity or by the decreasing redox potential [Sørensen, 1978*b*]. Nitrate reduction can also occur in deeper nitrate deficient sediments due to the action of burrowing animals [Sørensen, 1978*a*,*b*] and the burrows of benthic animals have been observed in Loch Eil sediment cores down to depths of 40 cm. In some continental shelf sediments, high rates of denitrification can occur at the sediment-water interface [Goering and Pamatat, 1971] whilst nitrate reduction is also thought to occur in reduced microniches [Sørensen, 1978*b*].

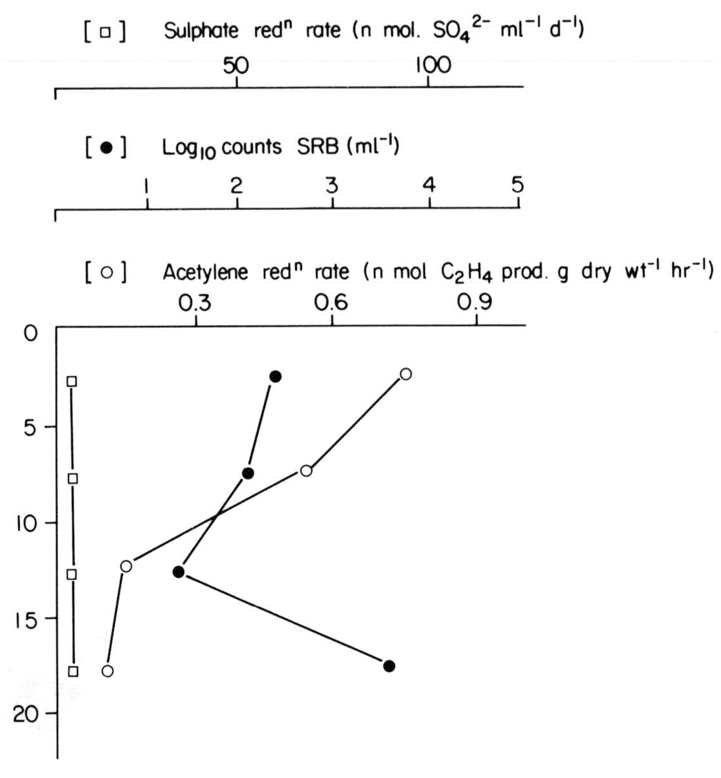

Fig. 6 Loch Creran CR1 - April 1980

Sulphate Reduction

As high concentrations of sulphate (approx. 30 mM) are normally found in sea water, the use of sulphate as terminal electron acceptor (by the sulphate reducing bacteria) is very often the predominant means of anaerobic oxidation in marine sediments. For example, in a New England salt marsh, sulphate reduction oxidized 1,800 g org-C. m^{-2} yr^{-1} whilst nitrate reduction only oxidized around 12 g org-C. m^{-2} yr^{-1} [Howarth and Teal, 1979]. In a model marine sediment system sulphate reduction accounted for up to 50% of the mineralization of organic detritus [Jørgensen and Fenchel, 1974]. Although sulphate reduction rates are limited by sulphate at concentrations below 2 to 10 mM [Howarth and Teal, 1979] such levels are rarely encountered in marine sediments due to the rapid exchange of sulphate from the overlying seawater to the sediment porewater. The sulphate concentration of porewater from the top 10 cm of a coastal sediment (the zone of greatest sulphate reduction) tends to be very similar to that of the overlying seawater

[Jørgensen, 1977*b*]. A profile of sulphate concentration with depth for Station E70 in Loch Eil is shown in Fig. 3 [F. Drake, unpub. data]. During this period (November, 1979) high rates of sulphate reduction were measured at this site (see below) reflected by the large decrease in sulphate concentration in the top 10 cm of the sediment core. However, even at a depth of 60 cm in the sediment the sulphate concentrations were in excess of 10 mM.

The use of sulphate as an inorganic terminal electron acceptor, proceeds according to the equation:

$$SO_4^{2-} + 4\ H_2O + H^+ \rightarrow HS^- + 4\ H_2O \quad \Delta G^{o1} = -36.4\ \text{kcal/mol}$$

[Thauer *et al.*, 1977]

This reaction, which requires an Eh of -150 to -200 mV, was until recently, thought to occur only in two genera of sulphate reducing bacteria, *Desulfovibrio* and *Desulfotomaculum* [Postgate, 1979]. Bacteria of these genera can use only a limited number of electron donors for sulphate reduction such as lactate, succinate, pyruvate, malate, fumarate, ethanol, formate, hydrogen, butanol and propanol [Wake *et al.*, 1977]. Sulphate reducing bacteria which could utilize acetate (except *Desulfotomaculum acetoxidans*) butyrate or propionate were thought not to exist. Recent work by Widdel [1979] has demonstrated the existence of organisms which can metabolize propionate and even-numbered lower and higher fatty acids (the genera *Desulfonema, Desulfovibrio* and *Desulfotomaculum*) and organisms which can utilize acetate, higher fatty acids and benzoate (the genera *Desulfobacter, Desulfococcus, Desulfonema, Desulfosarcina* and *Desulfotomaculum*). The activities of these newly discovered sulphate reducing bacteria may well be of importance in sediments such as Loch Eil where significant concentrations of acetate, butyrate and propionate have been measured [Miller *et al.*, 1979]. Counts of acetate and propionate utilizing sulphate reducing bacteria up to $10^3\ ml^{-1}$ have been measured in marine sediments [Laanbroek and Pfennig, 1981]. The results of a series of sulphate-reducing bacterial enrichments on a variety of electron donors from sediments from Station Loch Eil and from the North Eastern Atlantic Ocean are shown in Table 5. These data indicate the possibility of a wide range of electron donors for sulphate reduction being used in these sediments.

Direct measurements of sulphate reduction rates in marine sediments can be obtained by incubation of sediment samples with $^{35}SO_4^{2-}$. The labelled sulphide produced may be measured with a high degree of sensitivity allowing low rates to be measured over a short time period [Jørgensen, 1978]. Counts of sulphate reducing bacteria have traditionally been estimated by the use of lactate based media [see Pankhurst, 1971; Postgate, 1979], although media for the enumeration of acetate and

TABLE 5

Enrichments on different electron donors of sulphate reducing bacteria from Loch Eil and the N.E. Atlantic

Electron donor (0.1 w/v conc^n)	E70	A2	A3
Acetate	+	+	+
Benzoate	nd	-	-
Butanol	+	-	-
Butyrate	+	nd	+
Ethanol	+	+	+
Formate	+	nd	nd
Glucose	-	-	-
Glycerol	+	-	+
Hydrogen	+	+	+
Lactate	+	+	+
Propanol	+	-	+
Propionate	+	+	+
Pyruvate	+	+	+
Succinate	+	+	+
Valerate	-	-	-

+ growth; - no growth; nd not determined

propionate utilizing sulphate reducing bacteria have been described recently [Laanbroek and Pfennig, 1981]. Sulphate reduction rates and counts of lactate utilizing sulphate reducing bacteria have been measured in a variety of inshore, coastal and deep-sea sediments and are shown in Table 6. Highest rates of sulphate reduction were obtained in organically rich sediments with high inputs of organic material such as Station E70 in Loch Eil and Station D4 in the inner basin of Sullom Voe (Shetland Islands) with a stagnant body of overlying water in summer and high inputs of eroded peat particles [Pearson and Stanley, 1977]. Intermediate sulphate reduction rates were encountered in less organically rich sediments such as those found in Colla Firth (Shetland Islands), Station E24 in Loch Eil and Station CR2 in Loch Creran (N.W. Scotland). Sediment from Station CR1 in Loch Creran,

TABLE 6

Sulphate reduction rates and counts of lactate utilizing sulphate reducing bacteria in the top 5 cm of some marine sediments [Battersby, Leftley, Stanley and Brown, 1981]

Sampling station and depth	Sulphate reduction rate n mol SO_4^{2-} ml^{-1} d^{-1}	Sulphate reducing bacteria (ml^{-1})
DEEP SEA		
N.E. Atlantic A1 (4,920 m)	-	-
N.E. Atlantic A2 (2,880 m)	0.1	nd
COASTAL		
N.E. Atlantic A3 (158 m)	1.0	nd
INSHORE		
Loch Linnhe LY1 (47 m)	0.1	3.0×10^2
Loch Eil E24 (30 m)	15.0	1.2×10^3
E70 (49 m)	150	1.0×10^2
Loch Creran CR1 (13 m)	4.0	2.3×10^2
CR2 (18 m)	10.0	7.9×10^2
Sullom Voe D4 (44 m)	100	4.9×10^3
Colla Firth CF1 (24 m)	40	3.3×10^3

\- not detected nd not determined

although grossly contaminated by effluent from a nearby alginate factory, (% C of sediment = 20%), had a much lower activity of sulphate reduction than Stations E24 (% C = 10%) and CR2 (% C = 2%). Anaerobic oxidation in CR1 sediment appears to be dominated by methanogenesis [F. Drake, unpub.] and at present the reason for this is unclear. The lowest rates of sulphate reduction were measured in oxidized sediments of low organic content such as LY1, A3 on the western continental shelf and A2 in the abyssal Atlantic plain. No sulphate reduction was detected in

A1 sediment and attempted counts and enrichments for sulphate reducing bacteria yielded negative results. Counts of lactate utilizing sulphate reducing bacteria showed no significant correlation between numbers and rates as has been the case in other studies [Jørgensen, 1977*b*; Nedwell and Abram, 1978]. This may be due to the quality (i.e. decomposibility) of organic material reaching the sediment influencing the rate of sulphate reduction and not cell numbers, or to the presence of sulphate reducing bacteria other than lactate utilizers [Laanbroek and Pfennig, 1981].

Nitrogen Fixation

In many marine sediments the availability of nitrogen may be a limiting factor in the anaerobic breakdown of detritus. The occurrence of heterotrophic nitrogen fixing bacteria has been reported in a number of marine sediments [e.g. Pshenin, 1963; Herbert, 1975] and sulphate reducing bacteria of the genus *Desulfovibrio* implicated as the principle heterotrophic nitrogen fixing bacteria under marine conditions [Herbert, 1975; Blake and Leftley, 1977; Blake *et al.*, 1982]. Nitrogen fixation can be measured conveniently by the acetylene reduction test [Stewart *et al.*, 1967] in which sediment samples are incubated in the dark (to discount blue-green bacterial activity) in the presence of acetylene. Any ethylene produced (by the action of nitrogenase) may be estimated by gas chromatography to yield an indirect measurement of heterotrophic nitrogen fixation activity. Acetylene reduction rates in a number of marine sediments have been measured in parallel with estimations of sulphate reduction rate and counts of lactate utilizing sulphate reducing bacteria. Three profiles obtained are shown in Figs. 4, 5 and 6.

Sediments from LY1 showed a fairly constant rate of acetylene reduction with depth, together with low counts of sulphate reducing bacteria. The rate of sulphate reduction was very low and increased with depth, indicating the oxidized nature of the surface sediments with sulphate reduction probably occurring in reduced microniches within the surface sediments [see Jørgensen, 1977*a*]. In D4 sediments the maximum rates of acetylene and sulphate reduction and highest counts of sulphate reducing bacteria occurred in the surface sediments and decreased rapidly with depth. During the period of sampling (June) the overlaying water in the inner basin of Sullom Voe stagnated and was oxygen depleted [T.H. Pearson and S.O. Stanley, unpub.] enabling maximum rates of anaerobic decomposition to occur in the more organically enriched surface sediments. Sediments from station CR1 in Loch Creran showed low counts of sulphate reducing bacteria and a low constant sulphate reduction rate. Anaerobic breakdown of detritus within the sediment occurred mainly via methanogenesis [F. Drake, unpub.]. High acetylene reduction rates occurred in the surface sediment and decreased rapidly with depth, probably reflecting the decrease of oxidizable substrates with depth, which may restrict heterotrophic nitrogen fixation [see Stewart, 1969]. There was no clear relationship between counts of sulphate reducing bacteria, and more importantly, sulphate reduction rate and acetylene reduction rate. Due to the dependence of heterotrophic nitrogen fixation on sources of organic carbon [Stewart, 1969] it would seem reasonable to assume that if sulphate reducing bacteria were the predominant heterotrophic nitrogen fixers in the marine environment there would be a close correlation between rates of sulphate reduction and acetylene reduction. This appeared not to be the case in the sediments examined. Marijama *et al.* [1964] showed stimulation of acetylene reduction in various sediments (Japanese coast) by addition of sucrose and mannitol which cannot be utilized by sulphate reducing bacteria. Blake and Leftley [1977] also showed no correlation between counts of sulphate reducing bacteria and acetylene reduction rates. This suggests that in those marine sediments so far examined, sulphate reducing bacteria may not after all be the predominant heterotrophic nitrogen fixing bacteria.

References

Abram, J.W. and Nedwell, D.B. (1978). Inhibition of methanogenesis by sulphate reducing bacteria competing for transferred hydrogen. *Archives of Microbiology* **117**, 89-92.

Adams, C.A., Warnes, G.M. and Nicholas, D.J.D. (1971). A sulphite-dependent nitrate reductase from *Thiobacillus denitrificans*. *Biochimica et Biophysica Acta* **235**, 398-406.

Blake, D. and Leftley, J.W. (1977). Studies on anaerobic nitrogen fixation in the sediments of two Scottish sea-lochs. In "Biology of Benthic Organisms" (Eds. B.F. Keegan, P.O. Ceidigh and P.J.S. Boaden), pp.79-84. Pergamon Press, Oxford.

Blake, D., Leftley, J.W. and Brown, C.M. (1982). The bacterial flora and heterotrophic nitrification in sediments of Loch Eil. *Journal of Experimental Marine Biology and Ecology* **56**, 115-122.

Cole, J.A. and Brown, C.M. (1980). Nitrite reduction to ammonia by fermentative bacteria: a short circuit in the biological nitrogen cycle. *FEMS Microbiology Letters* **7**, 65-72.

Dunn, G.M., Wardell, J.N., Herbert, R.A. and Brown, C.M. (1980). Enrichment, enumeration and characterization of nitrate-reducing bacteria present in sediments in the River Tay estuary. *Proceedings of the Royal Society of Edinburgh* **78B**, 47-56.

Fenchel, T.M. and Jørgensen, B.B. (1977). Detritus carbon, nitrogen and sulphur. *Society for General Microbiology Quarterly* **6**, 7-8.

Goering, J.J. and Pamatat, M.M. (1971). Denitrification in sediments of the sea off Peru. *Invest. Resq.* **39**, 233-247.

Goman, G.A. (1973). Aerobic cellulose decomposition in the bottom of Lake Baikal. *Mikrobiologiya* **4**, 148-153.

Haddock, B.A. and Jones, C.W. (1977). Bacterial respiration. *Bacteriological Reviews* **41**, 47-99.

Herbert, R.A. (1975). Heterotrophic nitrogen fixation in shallow estuarine sediments. *Journal of Experimental Marine Biology and Ecology* **18**, 215-225.
Herbert, R.A., Dunn, G.M. and Brown, C.M. (1980). The physiology of nitrate dissimilatory bacteria from the Tay Estuary. *Proceedings of the Royal Society of Edinburgh* **78B**, 79-87.
Howarth, R.W. and Teal, J.M. (1979). Sulphate reduction in a New England salt marsh. *Limnology and Oceanography* **24**, 999, 1013.
Jones, J.G. and Simon, B.M. (1980). Decomposition processes in the profundal region of Blelham Tarn and the Lund tubes. *Journal of Ecology* **68**, 493-512.
Jones, J.G. and Simon, B.M. (1981). Differences in microbial decomposition processes in profundal and littoral lake sediments, with particular reference to the nitrogen cycle. *Journal of General Microbiology* **123**, 297-312.
Jørgensen, B.B. (1977*a*). Bacterial sulphate reduction within reduced microniches of oxidized marine sediments. *Marine Biology* **41**, 7-17.
Jørgensen, B.B. (1977*b*). The sulphur cycle of a coastal marine sediment (Limfjorden, Denmark). *Limnology and Oceanography* **22**, 814-832.
Jørgensen, B.B. (1978). A comparison of methods for the quantification of bacterial sulphate reduction in coastal marine sediment. I. Measurement with radiotracer techniques. *Geomicrobiology Journal* **1**, 11-27.
Jørgensen, B.B. and Fenchel, T. (1974). The sulphur cycle of a marine sediment model system. *Marine Biology* **24**, 189-201.
Koike, I. and Hattori, A. (1978). Denitrification and ammonia formation in anaerobic coastal sediments. *Applied and Environmental Microbiology* **35**, 278-282.
Laanbroek, M.J. and Pfennig, N. (1981). Oxidation of short-chain fatty acids by sulphate-reducing bacteria in freshwater and in marine sediments. *Archives of Microbiology* **128**, 330-335.
Marijama, Y., Suzuki, T. and Otobe, K. (1974). Nitrogen fixation in the marine environment: the effect of organic substrates on acetylene reduction. In "The Effect of Ocean Environment on Microbial Activities" (Eds. R.R. Colwell and R.Y. Morita), pp.341-353. University Park Press, Baltimore.
Martens, C.S. and Berner, P.A. (1974). Methane production in the interstitial waters of sulphate-depleted marine sediments. *Science* **185**, 1167-1169.
Miller, D., Brown, C.M., Pearson, T.H. and Stanley, S.O. (1979). Some biologically important low molecular weight organic acids in the sediments of Loch Eil. *Marine Biology* **50**, 375-383.
Nedwell, D.B. (1975). Inorganic nitrogen metabolism in a eutrophicated tropical mangrove estuary. *Water Research* **9**, 221-231.
Nedwell, D.B. and Abram, J.W. (1978). Bacterial sulphate reduction in relation to sulphur geochemistry in two contrasting areas of saltmarsh sediment. *Estuarine Coastal Marine Science* **6**, 341-351.
Pankhurst, E.S. (1971). The isolation and enumeration of sulphate-reducing bacteria. In "Isolation of Anaerobes" (Eds. D.A. Shapton and R.G. Board), pp.223-240. Academic Press, London.
Payne, W.J. (1973). Reduction of nitrogenous oxides by microorganisms. *Bacteriological Reviews* **37**, 409-452.
Pearson, T.H. (1982). The Loch Eil Project: assessment and synthesis, *Journal of Experimental Marine Biology and Ecology* **57**, 93-124.

Pearson, T.H. and Stanley, S.O. (1977). The benthic ecology of some Shetland Voes. In "Biology of Benthic Organisms" (Eds. B.F. Keegan, P.O. Ceidiah and P.J.S. Boaden), pp.503-512. Pergamon Press, Oxford.

Pearson, T.H. and Stanley, S.O. (1979). Comparative measurement of the redox potential of marine sediments as a rapid means of assessing the effect of organic pollution. *Marine Biology* **53**, 371-379.

Postgate, J.R. (1979). "The Sulphate-Reducing Bacteria" Cambridge University Press, Cambridge.

Pshenin, L.N. (1963). Distribution and ecology of *Azotobacter* in the Black Sea. In "Symposium on Marine Microbiology" (Ed. C.H. Oppenheimer), pp.383-391. Thomas, Illinois.

Sørensen, J. (1978*a*). Capacity for denitrification and reduction of nitrate to ammonia in a coastal marine sediment. *Applied and Environmental Microbiology* **35**, 301-305.

Sørensen, J. (1978*b*). Occurrence of nitric and nitrous oxides in a coastal sediment. *Applied and Environmental Microbiology* **36**, 809-813.

Sørensen, J., Jørgensen, B.B. and Revsbech, N.P. (1979). A comparison of oxygen, nitrate and sulphate respiration in coastal marine sediment. *Microbial Ecology* **5**, 105-115.

Stanley, S.O., Pearson, T.H. and Brown, C.M. (1978). Marine microbial ecosystems and the degradation of organic pollutants. In "The Oil Industry and Microbial Ecosystems" (Eds. K.W.A. Chater and M.J. Somerville), pp.60-79. Heyden and Son, London.

Stanley, S.O., Leftley, J., Miller, D. and Pearson, T.H. (1980). Chemical changes in the sediments of Loch Eil arising from the input of cellulose fibre. In "Analytical Techniques in Environmental Chemistry" (Ed. J. Albaiges), pp.409-418. Pergamon Press, Oxford.

Stewart, W.D.P. (1969). Biological and ecological aspects of nitrogen fixation by free living microorganisms. *Proceedings of the Royal Society of London, Series B* **132**, 367-388.

Stewart, W.D.P., Fitzgerald, G.P. and Burris, R.H. (1967). *In situ* studies on N_2 fixation using the acetylene reduction technique. *Proceedings of the National Academy of Sciences U.S.A.* **58**, 2071-2078.

Taylor, B.F., Campbell, W.L. and Chinoy, I. (1970). Anaerobic degradation of the benzene nucleus by a facultatively anaerobic microorganism. *Journal of Bacteriology* **102**, 430-437.

Taylor, B.F. and Heeb, M.J. (1972). The anaerobic degradation of aromatic compounds by a denitrifying bacterium. *Archiv. fur Mikrobiologie* **83**, 165-171.

Thauer, R.K., Jungermann, K. and Decker, K. (1977). Energy conservation in chemotrophic anaerobic bacteria. *Bacteriological Reviews* **41**, 100-180.

Traxler, R.W. and Bernard, J.M. (1969). The utilization of n-alkanes by *Pseudomonas aeruginosa* under conditions of anaerobiosis. Preliminary observation. *Int. Biodeterior. Bull.* **5**, 21-25.

Vance, I. (1977). Bacterial degradation of cellulose in marine sediments. Ph.D. Thesis. University of Dundee.

Vance, I., Stanley, S.O. and Brown, C.M. (1979). A microscopical investigation of the bacterial degradation of wood pulp in a simulated marine environment. *Journal of General Microbiology* **114**, 69-74.

Vance, I., Stanley, S.O. and Brown, C.M. (1982). Cellulose degrading bacteria in the sediments of Loch Eil and the Lynn of Lorne. *Journal of Experimental Marine Biology and Ecology* **56**, 267-278.

Wake, L.V., Christopher, R.K., Rickard, A.D., Anderson, J.E. and Ralph, B.J. (1977). A thermodynamic assessment of possible substrates for sulphate-reducing bacteria. *Australian Journal of Biological Sciences* **30**, 155-172.

Widdel, F. (1979). *Anaerober Abbau von Fettsauren und Benzoesaure durch neu isolierte Arten Sulfat-reduzierender Bakterien.* Doctoral thesis, University of Gottingen.

Zobell, C.E. (1938). Studies on the bacterial flora of marine bottom sediments. *Journal of Sedimentary Petrology* **8**, 10-18.

Chapter 7

INTERACTIONS BETWEEN BACTERIA AND BENTHIC INVERTEBRATES

J.C. FRY

Department of Applied Biology, University of Wales Institute of Science and Technology, King Edward VII Avenue, Cardiff, Wales, UK

Introduction

Sediments often contain large populations of both invertebrates and bacteria. Many benthic invertebrates burrow into the substratum and hence live in close association with bacteria which are present throughout sediments. Consequently it is logical that many interactions should have evolved between these two groups of organisms. Benthic invertebrates can be divided into three categories, these are the microbenthos or protozoans, the meiofauna which are microscopic metazoans, and the macrofauna or larger metazoans. Interactions between protozoans and bacteria have been discussed elsewhere [Coull, 1973; Fenchel, 1978; Fenchel, 1980] and although these studies have concentrated on feeding protozoans will not be considered here. The possible interactions between the metazoans and bacteria are numerous. Consequently specialized interactions involving, for example, parasitism, disease and symbiosis will not be considered. The remaining interactions (Fig. 1) involve the normal bacterial flora of the sediment and will be the subject of this chapter. Similar interactions probably occur in freshwater, estuarine and marine habitats but as the latter two habitats have been reviewed recently [Coull, 1973; Tietjen, 1979] freshwater sediments will receive most attention here.

Comparatively little work has been published about the interactions between benthic invertebrates and bacteria and the quantitative significance of the interactions is largely unknown [Tietjen, 1979]. Consequently this review will present the results of some calculations which attempt to indicate the magnitude of some of the interactions which occur. Although these calculations are often based on a minimum of information I hope they will make ecologists aware of the potential magnitude of the interactions and

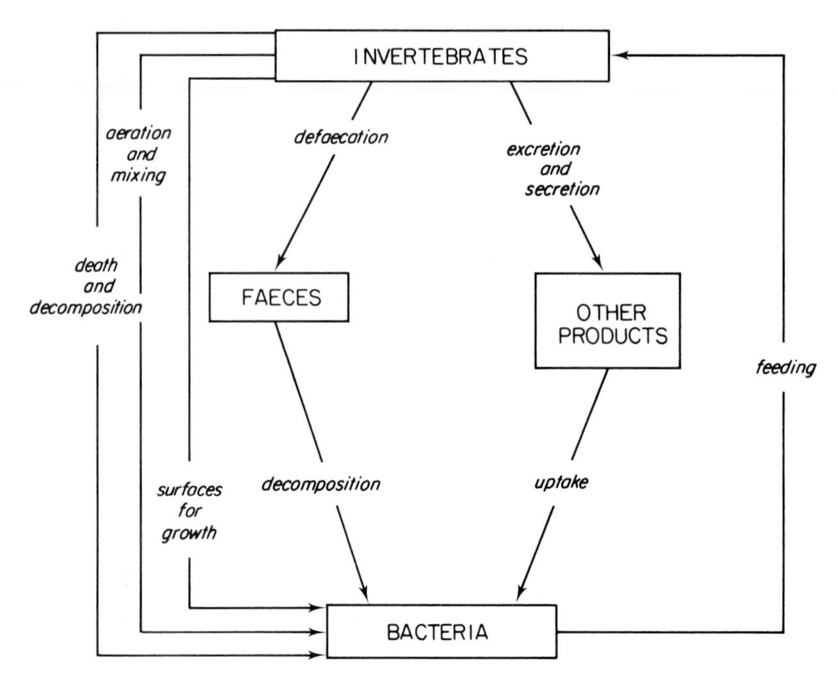

Fig. 1 Interactions occurring between invertebrates and bacteria in sediments.

help to stimulate research in this neglected area of ecology.

Invertebrate Data used in Calculations

Freshwater sediments contain a wide variety of invertebrates, but in most cases chironomids and oligochaetes predominate. Table 1, using data for three waters in South Wales and some other lakes, shows that this is true when calculated either from numerical data or by taking biomass into account. Consequently, for simplicity all calculations presented here will assume the sediments to contain significant numbers of these two groups of invertebrate only; and also the oligochaetes will be assumed to consist of tubificids and naidids only. The calculations will probably have general relevance to estuarine and marine habitats because oligochaetes often predominate in these habitats, particularly when organically enriched [Pearson and Rosenberg, 1978; Reish, 1979]. For the purposes of the calculations it is necessary to select sites which represent the range of numbers of chironomids, tubificids and naidids found in nature. Table 2 gives the numbers in these groups of invertebrates in the sites selected. They cover a full range of population densities

TABLE 1

Percentages of benthic macrofauna which are chironomids or oligochaetes in selected freshwater habitats

Sites	% of total macrofauna consisting of			Authors
	oligochaetes	chironomids	both	
River Taff*	75	10	85	Edwards *et al.* [1972]
Barry Reservoir*	34	29	63	Brooker and Edwards [1974]
Eglwys Nunydd Reservoir a*	14	83	97	Potter and Learner [1974]
Eglwys Nunydd Reservoir b+	11	79	90	Potter and Learner [1974]
Average from 7 North American oligotrophic lakes*	16	49	65	Wetzel [1975]
Average from 3 North American eutrophic lakes*	47	37	84	Wetzel [1975]

* Percentages calculated numerically

+ Percentages calculated on the basis of biomass

TABLE 2

Populations of chironomids, tubificid worms and naidid worms at selected sites in Great Britain*

Site		Number of invertebrates (m^{-2}) belonging to the			Authors
		Chironomidae	Tubificidae	Naididae	
River Taff	(min.)	116	72	280	Edwards *et al.* [1972]
River Taff	(av.)	1,749	8,515	5,540	Edwards *et al.* [1972]
River Taff	(max.)	9,818	122,000	34,240	Edwards *et al.* [1972]
Eglwys Nunydd Reservoir		28,800	4,812	-	Potter and Learner [1974]
Barry Reservoir		10,000	12,000	-	Brooker and Edwards [1974]
River Thames		-	5,740,000	-	Palmer [1968]

* Minimum, average and maximum populations are reported for the River Taff. Average values are reported for Eglwys Nunydd Reservoir and Barry Reservoir and the maximum values are reported for the River Thames.

from sparse populations such as the minimum value for the River Taff to the exceedingly dense populations reported for sediment at the low water mark of the River Thames at Battersea. Some represent the more usual population densities found in freshwaters, for example the population in Eglwys Nunydd Reservoir is very similar to that more recently reported for Lough Neagh [Carter, 1978] which is said to be similar to Loch Leven, Lake Erie and Lake Esrom.

The sizes of the invertebrates are also needed but as dimensions are rarely quoted in the literature these have to be calculated from individual animal biomasses. Although chironomid larvae are often large, up to between 1 and 3 mg dry weight [Johnson and Brinkhurst, 1971; Carter, 1978], mixed populations of smaller instars and species often occur. Hence an average chironomid biomass of 120 μg dry weight, which is calculated from the results reported by Potter and Learner [1974], will be used. These authors also give the dry weight as 14% of the fresh weight. Consequently, the average wet weight is about 860 μg, which is consistent with a chironomid of 0.4 mm diameter and 5 mm long. Average tubificid biomass also varies. From the results of Potter and Learner [1974] and Carter [1978] average values of 99 and 93 μg dry weight, respectively, can be calculated. However, sometimes tubificids reach an average of 250 μg dry weight [Palmer, 1968]. An average figure of 145 μg dry weight will be used here. As their dry weight is about 16% of the wet weight [Whitten and Goodnight, 1966; Appleby and Brinkhurst, 1970] the tubificid live weight will be 906 μg. From regression equations obtained in this laboratory which relate width and length to wet weight for tubificids [J.W. Densem, pers. comm.] this is equivalent to an animal of about 0.3 mm in diameter and 11 mm long. Sizes of naidids have been reported as generally less than 12 mm long and 0.15 mm in diameter [Harper *et al.*, 1981*a*] and most species are less than 10 mm in length [Learner *et al.*, 1978]. As sizes are often overstated in the literature unless measured directly from native populations, a size of 8 mm long and 0.15 mm in diameter will be assumed here. This corresponds to a naidid biomass of 183 μg wet weight and about 30 μg dry weight. The surface areas of the invertebrates to be used in the calculations are also needed. These are easily calculated assuming the animals to be cylindrical and are 10.5 mm^2 for the tubificid, 6.5 mm^2 for the chironomid and 3.8 mm^2 for the naidid.

Invertebrates as Surfaces for Bacterial Growth

The external surfaces of aquatic invertebrates are potentially an ideal habitat for bacterial colonization. Not only will such attached bacteria have all the advantages of the increased concentrations of nutrients, such as macromolecules and other organic materials associated with growth on any surface [Marshall, 1980], but they will have the advantages of the organic material secreted by the invertebrate on which they live. Consequently it is not surprising that bacteria are known to colonize the surfaces of many marine [Colwell and Liston, 1962; Boyle and Mitchell, 1978; Sochard *et al.*, 1979] and freshwater [Chatarpaul *et al.*, 1980; Harper *et al.*, 1981*a*] invertebrates. Taxonomic studies of these bacteria are rare. However Colwell and Liston [1962] reported that *Pseudomonas*, *Achromobacter* and *Flavobacterium* were the predominant bacteria on the eight marine and mainly benthic invertebrates examined. They also found little difference in bacterial flora between either the invertebrates or vertebrates examined. In another study, Sochard *et al.* [1979] found the bacterial flora on marine and estuarine copepods to be similar to that in the surrounding water. In this case *Vibrio* spp. predominated with *Pseudomonas*, *Cytophaga*/*Flavobacterium* and *Chromobacterium* also present. Numbers of bacteria on aquatic invertebrates are also rarely reported and where given are hard to compare. Up to 3×10^4 heterotrophic bacteria organism^{-1} have been found on marine copepods [Sochard *et al.*, 1979] by a viable counting procedure. In freshwater habitats, tubificid worms have been shown to have up to 2.1×10^4 denitrifying bacteria organism^{-1} and 1.7×10^5 nitrate reducing bacteria organism^{-1} growing on their surface [Chatarpaul *et al.*, 1980] whilst naidid worms had up to 2.96×10^5 bacteria organism^{-1} [Harper *et al.*, 1981*a*] when total numbers were estimated by the acridine orange epifluorescence technique. In the latter study epizoic bacteria were present on both the body wall and the chaetae of the animals (Fig. 2).

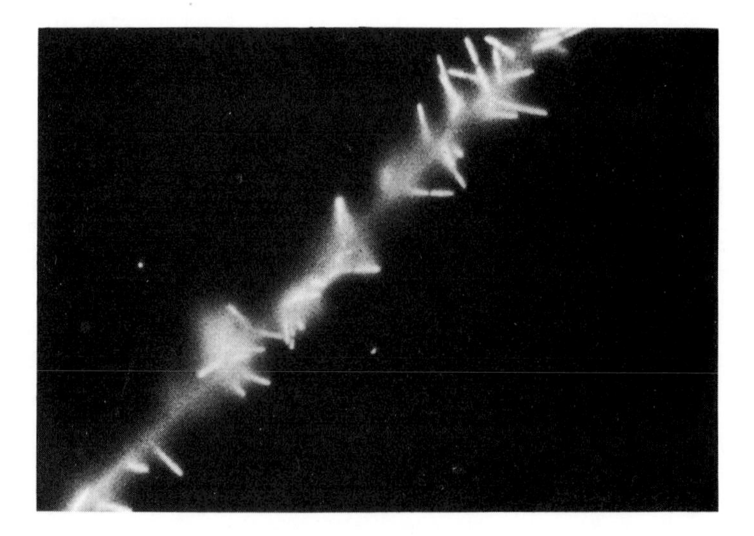

Fig. 2 Epizoic bacteria on a chaeta of *Nais variabilis*. A photograph taken using acridine orange stained worms viewed by epifluorescence microscopy as described by Harper *et al.* [1981*a*].

As little is known about the bacteria inhabiting the surfaces of benthic invertebrates it would be useful to calculate the number of bacteria on the surface of benthic invertebrate populations to see if they represented a significant proportion of the total sediment bacteria. The average total count of bacteria in the sediment of 16 Cumbrian lakes was about 5×10^{10} bacteria g dry weight^{-1} [Jones *et al.*, 1979] and maximum values in profundal sediments were only 1.2×10^{11} bacteria g dry weight^{-1} [Jones, 1980]. These values are similar to those obtained for the River North Tyne [Toscano and McLachlan, 1980] where total counts averaged about 1 to 2.5×10^{10} bacteria g dry weight^{-1}. Consequently as the total counts of bacteria in different sediments are reasonably similar, it is possible to use a single value in the calculations and 4×10^{10} bacteria g dry weight^{-1} of sediment will be used. Assuming sediment to be 60% water this is equivalent to 1.6×10^{10} bacteria ml^{-1} or 1.6×10^{14} bacteria m^{-2} for the top 1 cm of sediment which will be used here. As total counts of bacteria by epifluorescence microscopy have only been reported for *Nais*, a naidid, it is necessary to assume all naturally occurring invertebrates have a similar density of bacteria on their epidermis. This is about 3.0×10^5 bacteria worm^{-1} [Harper *et al.*, 1981*a*], hence the density of epizoic bacteria is 7.8×10^4 bacteria mm^{-2}. Table 3 gives the percentage of the sediment bacteria present on benthic invertebrates calculated using these assumptions. It can be seen that even for the dense populations of tubificids present in the River Thames benthos their epizoic bacteria are only 2.94% of those in the sediment. Hence it seems very unlikely that epizoic bacteria are important in the economy of the sediment.

Aeration and Mixing of Sediments

It has been demonstrated in a number of studies that many invertebrates mix the surfaces of sediments by both their burrowing and their feeding activities. For example, in laboratory experiments using irradiated scandium glass particles Edwards [1962] showed that tubificids were about one hundred times more effective in transferring particles from a depth of 5 cm to the surface of the sediment than were chironomid larvae. This difference was due to the feeding patterns of the two organisms, for tubificids feed well below the sediment surface while chironomids are surface feeders. In a later study at two sites in Lake Huron an invertebrate fauna, consisting mainly of amphipods and oligochaetes, was shown to mix the surface 3 cm at one site and 6 cm at the other [Krezoski *et al.*, 1978]. The increase in the mixed depth at the second site was explained by larger populations of the oligochaetes which mainly penetrated to 4 cm, instead of only 2 cm at the other site. McCall and Fisher [1980]

TABLE 3

Calculated numbers of epizoic bacteria on invertebrate populations and their percentage of the total bacterial population in the sediment

Site		Total number of epizoic bacteria ($\times 10^8 m^{-2}$) on			% of sediment bacteria that are epizoic
		chironomids	tubificids	naidids	
River Taff	(min.)	0.59	0.59	0.83	0.00013
	(av.)	0.20	69.7	16.4	0.0054
	(max)	8.87	999.0	101.0	0.0693
Eglyws Nunydd Reservoir		146.0	39.4	-	0.0116
Barry Reservoir		50.7	98.3	-	0.00931
River Thames		-	47000.0	-	2.94

have extensively investigated tubificid mixing activities in Lake Erie sediments, confirming and extending Edwards' [1962] earlier work. They have shown with a variety of techniques that the mixing is vertical and occurs because tubificids feed in the deeper layers of sediment and defaecate on the surface forming a pelletized layer 1 to 2 cm deep. Hence the depth of sediment is controlled by the depth of maximum feeding activity, which they show in microcosm experiments to be at about 5 cm; but they also acknowledge that this depth depends on the density of tubificids. They have also shown that the rate of tubificid mixing is greater than the sedimentation rate of new material and hence deposited material is immediately mixed into the sediment. This work has confirmed the dependence of the mixed depth on the depth of maximum feeding activity, an hypothesis which was originally suggested much earlier [Appleby and Brinkhurst, 1970], and also shows that the mixed depth is not purely dependent on the depth distribution of oligochaete numbers as many other workers have suggested [McCall and Fisher, 1980]. Similar processes have been shown to occur in marine sediments where macro-invertebrates transfer algae from the reduced zone to the oxidized surface layer and consequently increase the rate of bacterial mineralization in the sediment as a whole [Hylleberg and Henriksen, 1980]. This is one way in which these mixing activities can affect sediment bacteria.

It has been known for many years that burrowing benthic invertebrates, in particular chironomids, increase the oxygen uptake of sediment by more than the total of their own respiration and this was claimed to be due to the irrigation effects of the larvae while they ventilated their burrows [Edwards and Rolley, 1965]. There is substantial evidence to support this hypothesis. For example, the larvae of the chironomid *Chironomus riparius* have been shown, using lithium as a tracer, to more than double the exchange of interstitial water between sediment and overlying water and the exchange has also been shown to increase with increasing chironomid density [Edwards and Rolley, 1965]. Oligochaetes have also been shown to increase the circulation of water across the mud-water interface [Wood, 1975]. In this work, which used rhodamine B as a tracer, a substantial influx of water to the sediment was shown to occur. The magnitude of the influx depended upon the number of oligochaetes present and each worm was calculated to pump 9.5 to 15.0 μl of water into the sediment each hour. This pumping action by the macroinvertebrates oxidizes the sediment around their burrows. Such a phenomenon has been observed for many invertebrates, for example, *Chironomus plumosus* [Graneli, 1979]. Supporting evidence for these purely visual observations comes from early work [Edwards, 1958] which showed that increases in redox potential occurred in subsurface sediment when chironomid larvae were added. Recent work in which oxygen distribution in sediments has been measured with microelectrodes [Sørensen *et al.*, 1979; Revsbech *et al.*, 1980] has shown periodic oxygenation of sediment deeper than the surface layer, which is oxygenated by molecular diffusion and water turbulence. This was only explicable in terms of the burrowing activities of macrofauna and presents more positive evidence for the occurrence of an aerobic layer of sediment around invertebrate burrows than the previous observations. Despite this evidence, it is still suggested by some workers that oligochaetes do not pump water through burrows but that the observed water movements are due to increased permeability of the sediment caused by the burrows [McCall and Fisher, 1980].

The magnitude of the increased aeration of sediment is easily calculated; two approaches will be used here. Firstly, the increased surface area of aerobic sediment in invertebrate burrows will be calculated and secondly the increased oxygen uptake due to this increased surface area, presumably caused by irrigation of burrows by the invertebrates, will be estimated. In the first case it is assumed that the burrow is the diameter of the organism, is straight and 2 cm deep. This depth has been selected because, although oligochaetes and chironomids sometimes penetrate to 20 cm or more, most are present in the top 3 cm [Birtwell and Arthur, 1980] and by using 2 cm aeration effects will not be overestimated. Other workers have sometimes used different depths for this purpose, for example Graneli [1979] assumed a 30 cm U-shaped tube for *Chironomus plumosus* larvae to calculate that their burrows increased the surface area of sediment by 50%. It would appear necessary to subtract the cross-sectional area of the burrows to allow for the sediment surface removed by the animals, but in practice this is unnecessary as with the sizes of invertebrate used here this area is only 0.28% and 0.66% of the total burrow area for tubificids and chironomids respectively. The results of these calculations show (Table 4) that medium populations of benthic invertebrates such as those in Eglwys Nunydd reservoirs almost double the aerobic surface area but that the densest population in the River Thames increased the surface by an amazing 143 times.

Graneli [1979] has shown that *Chironomus plumosus* larvae increase sediment oxygen uptake by three to four times their own respiration, and this agrees with the factor of three for *Chironomus riparius* larvae calculable from earlier work [Edwards and Rolley, 1965]. Similar experiments have been used to show that tubificids increase sediment oxygen demand by the smaller factor of two [McCall and Fisher, 1980]. Consequently, an intermediate factor of three will be used for the next calculations. The respiration of the invertebrates needs to be calculated and this can be done assuming their oxygen uptake to be 70 μg mg dry weight^{-1} day^{-1} which is approximately the mean of the values reported by Johnson and

TABLE 4

Increases in the surface area of aerobic sediment due to the burrows of tubificids and chironomids

Site		Extra surface area (cm^2 m^{-2}) due to burrows of		% extra surface area due to burrows
		tubificids	chironomids	
River Taff	(min.)	18	22	0.4
	(av.)	2,129	332	24.6
	(max.)	30,500	1,865	324
Eglwys Nunydd Reservoir		1,203	5,472	66.8
Barry Reservoir		3,000	1,900	49.0
River Thames		1,435,000	-	1,435

Brinkhurst [1971] for chironomids and tubificids. The sediment oxygen uptake without invertebrates will be taken as 1500 mg m^{-2} day^{-1}. This is approximately the average value reported for Blelham Tarn sediment [Jones, 1976] and agrees closely with the figure of 1800 mg m^{-2} day^{-1} calculable from the results of Edwards and Rolley [1965], assuming that 40% of their sediment respiration was due to chironomids. This last assumption is stated in their paper and supported by recent work [Ripley, 1980]. The results of the calculations, based on these assumptions, show the percentage increase in oxygen uptake due to invertebrate burrowing activity (Table 5). They show a similar pattern to the increases in surface area, with the densest populations increasing oxygen uptake very greatly but the sparsest populations hardly affecting the sediment oxygen consumption. Sometimes tubificids share burrows (Fig. 3) and if this occurs frequently at high population densities the results of these calculations would be overestimates. Similarly it is not known whether sediment oxygen uptake is increased by invertebrates at the highest recorded population densities by the same amount as at lower population densities. This is a possible source of error as the enhancement of sediment oxygen consumption has only been measured at population

Fig. 3 Three individuals of the tubificid *Branchiura sowerbyi* occupying a single burrow in washed sand.

TABLE 5

Increases in the oxygen consumption of sediment due to the burrowing activities of tubificids and chironomids

Site		Extra oxygen uptake (mg m^{-2} day^{-1}) due to burrows of		% increase in oxygen uptake due to burrows
		tubificids	chironomids	
River Taff	(min.)	2	3	0.34
	(av.)	259	44	20.2
	(max)	3,715	247	264
Eglwys Nunydd Reservoir		147	726	58.2
Barry Reservoir		365	252	41.1
River Thames		174,783	-	11,652

densities of about 1% of the highest used in these calculations [Graneli, 1979; McCall and Fisher, 1980].

Any bacterial process dependant on oxygen is likely to be enhanced in sediments with dense populations of invertebrates. This is because large increases in both aerobic surface area and oxygen uptake are apparently possible in these sediments. Decomposition is one example of an oxygen dependant process, as aerobic decomposition by bacteria is much more rapid than the anaerobic process [Jørgensen, 1980]. Consequently, overall rates of decomposition should be greatly enhanced in sediments with large invertebrate populations [Fenchel, 1972; Pomeroy, 1979; Poole and Wildish, 1979].

Increases in other types of bacterial activity caused by benthic invertebrates have also been reported, with nitrogen and phosphorus metabolism being the most extensively investigated. Jacobsen [1977] showed little effect of macrofauna on the flux of nitrogen and phosphorus between sediment and water, but other workers have shown increased transport of both elements out of the sediment due to invertebrate activity in freshwater [Edwards, 1958; Chatarpaul *et al.*, 1979] and marine [Hylleberg and Henriksen, 1980, Henriksen *et al.*, 1980] habitats. This increased transport from sediment is due to both excretion by the animals and direct diffusion from the increased surface area of aerobic sediment in the burrows. Nitrate and ammonia contribute to the nitrogen flux and the nitrate flux is controlled by the balance of bacterial nitrification and denitrification in the sediment. The rate of both these processes is enhanced by the macrofauna, nitrification by about 240% and denitrification by about 180%. However, these increases were obtained in experiments using only 14,000 tubificids m^{-2} which is only in the mid range of the populations considered here. Consequently, much greater increases can be expected in the densest populations.

Microbial Interactions with Faeces

Many benthic invertebrates produce faeces which are relatively free of bacteria compared with the food they ingest [Wavre and Brinkhurst, 1971; Chua and Brinkhurst, 1973; Hargrave, 1976], and fresh faeces also have fewer species of bacteria associated with them [Brinkhurst and Chua, 1969; Chua and Brinkhurst, 1973]. However the faecal pellets of detritivores can be colonized by aquatic bacteria because little digestion of the ingested food takes place in the invertebrate gut [Hargrave, 1976] and detritus normally supports a large bacterial population [Fenchel, 1972]. Consequently, faeces provide ideal surfaces for bacterial growth once released into the sediment: such growth has been demonstrated by carbon and nitrogen analysis [Newell, 1965] and more directly by estimates of microbial biomass [Lopez *et al.*, 1977]. The bacterial flora develops rapidly on faeces, the bacterial cells being much larger than free-living bacteria [Pomeroy, 1979] and increasing in number by 180% in the first 1.5 days [Lopez *et al.*, 1977]. This increase is accompanied by a rapid increase in the rate of oxygen uptake of the faeces which reaches a maximum within 2 to 3 days. Thereafter oxygen consumption decreases just as rapidly to a low and stable level similar to that of detritus before ingestion [Hargrave, 1976]. Reworking of the sediment by feeding invertebrates [McCall and Fisher, 1980] could consequently result in periods of bacterial growth and high activity, alternating with periods of low activity. Thus in heavily reworked sediments, oxygen uptake could be increased by this cycle of events.

The potential magnitude of the increase in oxygen uptake can be calculated approximately from the data presented by Hargrave [1976] who reviews the importance of the faeces produced by a wide variety of benthic invertebrates. Hargrave [1976] standardizes units by expressing oxygen uptake per unit area of faeces, and faecal production as the surface area of faeces produced. This is also a useful standardization to use here because faecal oxygen uptake is better related to the surface area of faeces rather than its mass. The average rate of faecal production by the typical naidid, tubificid and chironomid can be calculated as 0.39, 0.798 and 0.725 mm^2 h^{-1} respectively, because Hargrave [1976] shows that individual biomass is related to faecal production. It is also necessary to assume an oxygen uptake rate for the faeces, 0.4 μg cm^{-2} h^{-1} is assumed here because this is the rate given for *Chironomus* feeding on lake sediment. In order to simplify the calculation, a decay curve has not been fitted to the rate of faecal oxygen uptake, but I have assumed that faecal material takes up oxygen at a constant rate for 24 h and then immediately stops consuming oxygen. The results in Table 6 have been obtained using these assumptions and show that in most cases the percentage oxygen uptake due to faeces is rather low. The highest value by far is that for the dense populations of tubificids in the River Thames and the 703.5% increase in oxygen uptake due to faeces calculated for this site appears very high. However, when calculated as a percentage of the oxygen uptake increase due to invertebrate burrows (Table 5), the value for the River Thames site is very close to the average value of 7.8% for all sites considered. Consequently, oxygen uptake due to faeces seems to represent only a small proportion of the total enhancement of sediment oxygen uptake caused by invertebrates.

Soluble Products from Invertebrates

Although it is well known that invertebrates release a range of dissolved organic materials by secretion through the epidermis and excretion, little detail is known of

TABLE 6

Increases in the oxygen consumption of sediment due to invertebrate faeces

Site		Extra oxygen uptake (mg m^{-2} day^{-1}) due to faeces of			% increase in oxygen uptake due to faeces
		naidids	tubificids	chironomids	
River Taff	(min.)	0.25	0.13	0.19	0.038
	(av.)	4.97	15.65	2.93	1.57
	(max.)	30.77	224.3	16.39	18.10
Eglwys Nunydd Reservoir		-	8.86	48.0	3.79
Barry Reservoir		-	22.06	16.73	2.58
River Thames		-	10552	-	703.5

these processes [Jørgensen, 1976]. It is likely, however, that much of this will be low molecular weight material and it is well known that bacteria remove such material very rapidly from aquatic habitats [Fenchel and Jørgensen, 1977; Pomeroy, 1979]. Consequently, it is probable that sediment bacteria rely in part on benthic invertebrates for their organic substrates but the magnitude of this dependence cannot at present be estimated.

Bacteria as Invertebrate Food

Many invertebrates living in sediment are detritivores and ingest large amounts of sediment in an indiscriminate manner [Fenchel and Jørgensen, 1977], although the maximum size of particles they can ingest will depend to some extent on the nature and size of their mouth and other organs or appendages used for ingestion. They are, however, very inefficient at utilizing the sediment as food. Consequently assimilation efficiencies for this diet are low and rarely exceed 50% [Monakov, 1972]. Berrie [1976] tabulates ten studies which have reported detritivore assimilation efficiencies which range from 2% to 44% with a mean of 16%; more recent work confirms these results and of particular interest here is the value of 4.1% recorded for two common tubificids [Brinkhurst and Austin, 1979]. It now appears generally accepted that these assimilation efficiencies are low because these detritivores only utilize the bacterial part of the food which they assimilate at high efficiency: for example, in two studies with purely bacterial food, assimilation efficiencies ranged from 60% to 97% and had an average value of 80% [Berrie, 1976].

The evidence that detritivores feed mainly on the bacterial part of their food is derived principally from studies with freshwater invertebrates, which show that most ingested bacteria are killed during passage through the gut. In larger organisms the evidence was obtained by counting bacteria in different parts of the dissected gut [Baker and Bradnam, 1976] or in faeces [Chua and Brinkhurst, 1973], but with smaller organisms, such as naidids, less direct techniques were used [Harper *et al.*, 1981*a*]. It has also been shown that bacterial death in the gut is a precursor of digestion and assimilation, because tubificids can take up and assimilate into their tissues radioactive phosphorus from [^{32}P]-labelled bacteria but not from labelled sediment [Whitten and Goodnight, 1969]. Similarly, a naidid has been shown to assimilate material from [^{3}H]-labelled bacteria [Harper *et al.*, 1981*b*]. Such assimilation must be due to digestion rather than adsorption of products secreted from the bacteria, as some bacteria fed to the naidid were labelled with [^{3}H]-thymidine. This was incorporated into the bacterial deoxyribonucleic acid which retained 95% of the label during the entire feeding period so was not secreted.

Conversely other work has shown that not all benthic invertebrates rely solely on bacteria as a source of food but can utilize dissolved organic matter directly. Some marine invertebrates, such as polychaetes, are thought to assimilate dissolved organic matter directly from the organically rich sediments in which they live [Jørgensen, 1976]. Similarly a nematode fed both [^{14}C]-glucose labelled bacteria and [^{14}C]-glucose in solution only took up label from solution [Lopez *et al.*, 1979]. However, direct assimilation of dissolved material seems a rarer feeding mechanism than bacterial digestion.

It is most unlikely that benthic detritivores selectively ingest bacteria because no differences in the range of bacteria found in food and gut contents has been found for either tubificids [Brinkhurst and Chua, 1969] or a naidid [Harper *et al.*, 1981*a*]. The occurrence of selective digestion is less clear. Baker and Bradnam [1976] found no differences in the bacterial composition of fore and hind guts of *Chironomus* and *Simulium* larvae, both of which digest bacteria, and so concluded that digestion was not selective. This hypothesis is supported by other work which has shown that *Simulium venustum* digested Gram positive and negative bacteria [Fredeen, 1964] and that the snail, *Planorbis contortus*, preferred a mixture of these types of bacteria on which to grow [Calow, 1974]. In some other invertebrates selective digestion has been shown to occur. For example, Brinkhurst and Chua [1969] showed that three species of tubificid each had only a single bacterial species in their gut after starvation for a week, indicating a different pattern of selective digestion by each species during starvation. As starvation is a rather artificial technique to use for organisms that probably feed continuously, another study, in which numbers of some different bacterial species in the food and faeces of the same three tubificids, were enumerated, probably gives more reliable results. This showed that *Flavobacterium* was the only type of bacterium which appeared to be digested by the worms as the numbers of this organism, which was the most abundant in the sediment, were fifteen times lower in the faeces than sediment [Wavre and Brinkhurst, 1971].

The distribution of benthic invertebrates might to some extent be controlled by their bacterial food. This has been demonstrated in the River North Tyne where the distribution of chironomid larvae was related closely to the amount of particulate material sedimented on the river bed, except in the peaty headwaters [Toscano and McLachlan, 1980]. In this region of the river there was an abundance of fine sediment but few larvae. Total counts of bacteria in sediment, by epifluorescence microscopy, showed these sediments to be unusually devoid of attached bacteria. Hence, the larval distribution was related more closely to the distribution of the attached bacteria than to the sediment itself. This work was supported by laboratory work which showed that *Chironomus lugubris* larvae preferred peat incubated with bacteria and fungi, rather than sterile peat [McLachlan and Dickinson, 1977].

Most studies on the feeding of benthic invertebrates on bacteria have involved organisms feeding on fine particulate material but some work on invertebrates which ingest larger items such as leaf fragments has been carried out. The water hoglouse, *Asellus aquaticus*, which typically inhabits small streams and the littoral region of lakes, has been shown to grow well on a variety of bacteria such as *Sphaerotilus natans*, *Actinoplanes* sp., *Micromonospora* sp. and *Streptomyces* spp., when these have been presented individually as sole food sources. *Asellus* also grows well on mixed cultures of tap-water bacteria and on bacteria epiphytic upon leaves of *Elodea canadensis* (Canadian pond weed) and *Quercus* spp. (oak) [Marcus *et al.*, 1978; Willoughby and Marcus, 1979]. The growth rate of *A. aquaticus* was slower on these bacteria and on epiphyte-covered leaves than on clean leaves of *E. canadensis* [Marcus *et al.*, 1978]: also faecal matter and lake sediment supported growth less rapidly than the other foods. Similar results have been reported for the chironomid, *Paratendipes albimanus*, which showed increased growth rates when fed leaves with a substantial community of epiphytic microorganisms than when fed faeces or natural detritus [Ward and Cummins, 1979].

Some invertebrates, particularly meiofauna, have specialized relationships with bacteria. This is reflected by the observation that some meiofauna are not only attracted to sediment particles coated in organic or bacterial films but are often preferentially attracted to sands coated with a bacterial community dominated by their preferred species [Coull, 1973]. Some invertebrates are thought specifically to encourage the growth of bacteria within sediment so they are secured a supply of bacterial food. For instance, some nematodes do this by secreting mucus on which bacteria grow and the animals then browse the mucus enriched with bacteria [Riemann and Schrage, 1978]. In the oncholaimid nematodes adults swallow detritus that remains undigested and which when defaecated forms agglutinations on which bacteria grow. Their eggs are laid in the agglutinations and the young feed by absorbing the organic matter secreted by the rich bacterial flora close by [Lopez *et al.*, 1979].

Most of the work discussed so far has been qualitative and only one study seems to have been undertaken specifically to determine the proportions of the natural microbial population grazed by invertebrates. This study was undertaken on an arctic tundra pond [Fenchel, 1975] and showed that almost 50% of the microbial production was consumed by invertebrates. Of this consumption, about two thirds was due to macroinvertebrates, protozoa and meiofauna accounting for equal proportions of the remainder.

The grazing activities of benthic invertebrates are believed to increase the growth rate and activity of sediment bacteria. Despite this often-quoted belief, little evidence for it exists in the literature. It has, however, been shown that protozoa increase the rate of hay decomposition in a mixed microbial culture and polysaccharide degradation rates are thought to be enhanced by macrofauna [Fenchel and Jørgensen, 1977; Poole and Wildish, 1979]. Bacterial growth rates are also probably increased under grazing pressure because microbes are removed from detritus by grazing and faecal material or grazed detritus which are free of microbes should be able to support more bacterial growth (the evidence and rationale for this has already been discussed in the previous section). A continuous alternation of growth and removal by feeding will increase growth rates of bacteria. Accordingly, Gerlach [1978] calculated for a model subtidal silty sand marine sediment at 30 m depth that 21 bacterial generations $year^{-1}$ were enough to support the subsurface bacteria-feeding fauna. This is only equivalent to a generation time of about 420 h, so represents a very slow growth rate.

Calculation of the growth rate of sediment bacteria that might be enforced by grazing pressure is a useful way of quantifying the effect of invertebrate grazing on bacteria and is undertaken here for the sites already used in other calculations. To make these estimates, the number of bacteria ingested by the typical naidid, tubificid and chironomid per unit of time must be calculated. This can be done if the gut volume, the gut retention time and the density of bacteria ingested are known. Unfortunately gut volumes are rarely quoted for benthic invertebrates. Consequently the volume of 0.05 mm^3 stated [Harper *et al.*, 1981*a*] for the naidid gut is used and values are calculated from this for tubificids and chironomids by assuming that gut volume changes in direct proportion to the dry weight of the organism. Using the average biomasses justified earlier, average gut volumes of 0.2 mm^3 and 0.24 mm^3 were obtained for chironomids and tubificids respectively. A gut retention time of 40 min for the naidids [Harper *et al.*, 1981*a*] was used and a value of 30 min was assumed for the tubificids and chironomids because this is an average value obtained from several macroinvertebrates [Berrie, 1976]. From this information the number of bacteria ingested per day was calculated making two assumptions. The first was that all the invertebrates had the same density of bacteria in their gut as the naidids; equivalent to 2.4×10^4 bacteria mm^{-3} which was calculated from the average number of viable bacteria in the naidid gut [Harper *et al.*, 1981*a*]. The second assumption was that 82% of the ingested bacteria were killed in the gut [Harper *et al.*, 1981*a*]. This meant that the average naidid, tubificid and chironomid ingested 2.4×10^5, 1.6×10^6 and 1.3×10^6 bacteria day^{-1} when estimated by viable counting on casein-peptone-starch (CPS) medium.

All the chironomids and oligochaetes were assumed to feed on sediment. This is undoubtedly true of oligochaetes but not entirely true for chironomids. However, Walshe [1951] has shown that although some benthic chironomids are filter feeders, utilizing suspending material in the water column, most ingest sediment particles either at the surface or deeper within the sediment. The bacteria ingested per unit area were then calculated (Table 7). To calculate the percentage of the sediment bacteria ingested only the top 2 cm of sediment was considered, which is where most invertebrates and bacteria would be in a typical sediment. This layer contains 2 x 10^{11} viable bacteria m^{-2}, when enumerated on CPS medium [Harper *et al.*, 1981*a*]. This population of sediment bacteria seems reasonable to use because it is based on the figure of 2.5 x 10^7 bacteria g dry weight^{-1} of sediment obtained from the River Ely [Fry and Staples, 1976] which was midway between the average populations of about 1 x 10^7 bacteria g^{-1} obtained for 16 Cumbrian lakes [Jones *et al.*, 1979] and 5 x 10^7 bacteria g^{-1} from Barry Reservoir [Fry *et al.*, 1973], all of which were estimated with similar techniques. The enforced mean generation time can then simply be calculated as the time taken for the invertebrates to ingest 69% of the sediment bacteria. This is a reasonable approximation because most bacteria are killed in the invertebrate gut. The results (Table 7) show that for most sites the enforced growth rate is close to the average values of between 3 and 51 h reported for the mean generation times of stream bacteria [Bott, 1975]. In the least densely populated River Taff site it is unlikely that invertebrate grazing has any effect on the natural bacterial growth rates. In the densely populated River Thames site the very short enforced mean generation time of about 22 min appears to represent a most unlikely rate of growth in a natural sediment (see Conclusions for further details).

Death and Decomposition of Invertebrates

The only benthic invertebrates that will not die in the sediment will be those insect larvae, such as chironomids, that develop into the emerging adults and these adults represent only a small proportion of the benthic population [Potter and Learner, 1974]. Although Titmus and Badcock [1980] report that between 30% and 50% of the benthic production of chironomids emerge as adults these figures are likely to be overestimates as these workers only use large (> 6 mm) larvae to estimate production. Consequently almost all the invertebrate biomass produced in the sediment will eventually be available as nutrients for bacteria during decomposition. Although this interaction between invertebrates and bacteria must occur in sediments it has received little attention in the literature [Tietjen, 1979]. Consequently some calculations

TABLE 7

Numbers of viable bacteria ingested by benthic invertebrates and the mean generation time enforced on sediment bacteria by grazing pressure

Site		Number of bacteria ingested (x 10^9 day^{-1} m^{-2}) by			% of sediment bacteria ingested (day^{-1})	Enforced mean generation time (h)
		naidids	tubificids	chironomids		
River Taff	(min.)	0.067	0.115	0.151	0.18	9240.0
	(av.)	1.33	13.6	2.27	8.6	193.0
	(max.)	8.22	195.0	12.8	108.0	15.4
Eglwys Nunydd Reservoir		-	7.7	37.4	22.6	73.6
Barry Reservoir		-	19.2	13.0	16.1	103.0
River Thames		-	9180.0	-	4590.0	0.36

TABLE 8

Production of benthic naidids, tubificids and chironomids and the production of bacteria expected from the decomposition of these invertebrates

Site		Total invertebrate production ($g\ m^{-2}\ y^{-1}$)	Bacterial production expected from invertebrate decomposition ($g\ m^{-2}\ y^{-1}$)	Total number of bacteria produced ($\times 10^{12}\ m^{-2}\ y^{-1}$)	Number of generations required for the bacterial production (y^{-1})
River Taff	(min.)	0.18	0.11	0.55	0.00172
River Taff	(av.)	8.85	5.31	26.6	0.083
River Taff	(max.)	109.43	65.66	328.0	1.03
Eglwys Nunydd Reservoir		22.84	13.70	68.5	0.214
Barry Reservoir		16.17	9.70	48.5	0.152
River Thames		4577.0	2746.2	13700.0	42.81

will be attempted here which attempt to show the magnitude of this interaction (Table 8).

The invertebrate production has been estimated from the product of the population density, individual biomass and the annual P/B ratio. All except the latter factor have been discussed earlier and a P/B ratio of 5.5 has been assumed here which is the average value for chironomids and oligochaetes given in a recent review of secondary production [Waters, 1977]. This gives reasonable values for invertebrate production, as the calculated value for Eglwys Nunydd reservoir is about 90% of the actual chironomid and oligochaete production [Potter and Learner, 1974]. The bacterial production possible during invertebrate decomposition was then calculated assuming an assimilation efficiency of 60% [Calow, 1977; Pomeroy, 1979] in both culture and in natural situations. To calculate the number of bacteria produced, the biomass of an individual bacterium is assumed to be 2×10^{-13} g, a figure similar to those used previously for similar types of calculation [Baker and Bradnam, 1976; Gerlach, 1978]. The number of bacterial generations per year which would be necessary to support this number of bacteria is then calculated using the average values for total numbers of bacteria given in the third section of this review. The top 2 cm was again used in these calculations so the results were comparable with those calculated in the preceding section.

The results of these calculations (Table 8) show that even at the highest density of invertebrates in the River Thames only about 43 generations of bacteria would be needed every year to decompose the total production of invertebrates. This is equivalent to an enforced mean generation time of only 205 h which represents considerably slower growth than that enforced by grazing pressure (Table 7). Consequently it seems unlikely that decomposition of dead invertebrates would contribute much to the overall bacterial productivity of sediments and that most of the organic carbon needed to support the growth rates predicted by grazing pressure must come from other sources, such as the detritus ingested by the invertebrates.

Conclusions

It is clear from the preceding sections of this review that although a body of literature exists which is relevant to the interactions between invertebrates and bacteria in sediments most of the information is qualitative, little being of a quantitative nature. From the calculations presented here it appears that epizoic bacteria do not make up a substantial proportion of the benthic bacterial community and that dead invertebrates are unlikely to be an important source of bacterial nutrients. However, in some other interactions benthic invertebrates can have a substantial effect on the sediment bacteria.

This is particularly likely to arise from the invertebrates burrowing activities, which greatly increase oxygen in the sediment and hence aerobic bacterial activities, or from grazing which appears to stimulate bacterial growth. Oxygen uptake due to bacteria growing on faecal pellets is, however, unlikely to contribute greatly to the total oxygen uptake of the sediment community. Bacteria are also important to the benthic invertebrate community. In particular they are the main food source for many detritivores and by colonizing faeces allow these invertebrates to reutilize sediment many times; this probably allows large populations of macrofauna to develop in some sediments, relatively unrestricted by food supply.

The River Thames site used in these calculations has a far denser population of invertebrates than any of the other sites used and it might appear to give rather unlikely results to many of the calculations. However, this is not a unique site as Caspers [1980] reports that Hamburg harbour has 800,000 tubificids m^{-2} and that these are large organisms which average 0.73 mm in diameter. By using regression equations which relate width, length and biomass, this population of tubificids must have a biomass of about 1560 g dry weight m^{-2} which is greater than that for the River Thames site which has 1435 g dry weight m^{-2} [Palmer, 1968]. Several other reports exist in the literature for tubificid densities between the largest population on the River Taff, which has a biomass of only about 20 g m^{-2}, and the River Thames population. Two examples are the River Thames at Erith (240 g dry weight m^{-2}; Birtwell and Arthur, 1980) and Toronto harbour (98.5 g dry weight m^{-2}; Brinkhurst, 1970). The tubificid densities reported for the River Thames and Hamburg harbour must represent populations very close to the theoretical maximum density because the proportion of the sediment surface area occupied by the worms, which are all assumed to be in a vertical position in the sediment, can be calculated as 72% and 34% respectively which leaves hardly any space between the worms [Caspers, 1980].

Chironomids do not appear to reach such high population densities as oligochaetes in sediments. However, *Chironomus riparius* densities of 100,000 m^{-2} in a sewage polluted stream [Edwards, 1957] and 132,000 m^{-2} in a thermally and waste-water polluted urban channel [Koehn and Frank, 1980] have been reported. Individual biomass data obtained in this laboratory for this species in a polluted river suggest that these densities approximate to biomass of 32 and 42 g dry weight m^{-2}. Although this is considerably lower than the biomass of the densest tubificid populations, they represent a much larger biomass than that of the densest River Taff site. Thus the densest populations of chironomids will probably also have a marked effect on bacteria in sediments.

Although the conclusions presented here are based only upon calculations which have many assumptions and use average values obtained from a disparate set of sites and hanitats, they will, I believe, form a basis for discussion and further research in the absence of more reliable information.

Acknowledgements

I am particularly grateful to my colleagues in UWIST who undertake research on the ecology of freshwater invertebrates for the background information they have given me and the many useful discussions I have had with them. In particular, I thank Mr J.W. Densem for allowing me to use his photograph of *Branchiura sowerbyi*, Dr M.A. Learner and Mr M. Holloway. I am also grateful to Dr R. Harper for allowing me to use his photograph of bacteria on the chaeta of *Nais variabilis* and to Professor R.W. Edwards for his helpful criticism of the manuscript.

References

Appleby, A.G. and Brinkhurst, R.O. (1970). Defaecation rate of three tubificid oligochaetes found in the sediment of Toronto Harbour, Ontario. *Journal of the Fisheries Research Board of Canada* **27**, 1971-1982.

Baker, J.H. and Bradnam, L.A. (1976). The role of bacteria in the nutrition of aquatic detritivores. *Oecologia* (Berlin) **24**, 95-104.

Berrie, A.D. (1976). Detritus, microorganisms and animals in fresh water. In "The Role of Terrestrial and Aquatic Organisms in Decomposition Processes" (Eds. J.M. Anderson and A. Macfadyen), pp.323-338. Blackwell Scientific Publications, Oxford.

Birtwell, I.K. and Arthur, D.R. (1980). The ecology of tubificids in the Thames estuary with particular reference to *Tubifex costatus* (Claparède). In "Aquatic Oligochaete Biology" (Eds. R.O. Brinkhurst and D.G. Cook), pp.331-381. Plenum Press, New York.

Bott, T.L. (1975). Bacterial growth rates and temperature optima in a stream with a fluctuating thermal regime. *Limnology and Oceanography* **20**, 191-197.

Boyle, P.J. and Mitchell, R. (1978). Absence of microorganisms in crustacean digestive tracts. *Science* **200**, 1157-1159.

Brinkhurst, R.O. (1970). Distribution and abundance of tubificid (Oligochaeta) species in Toronto Harbour, Lake Ontario. *Journal of the Fisheries Research Board of Canada* **27**, 1961-1969.

Brinkhurst, R.O. and Austin, M.J. (1979). Assimilation by aquatic oligochaeta. *Internationale Revue der gesamten Hydrobiologie und Hydrographie* **64**, 245-250.

Brinkhurst, R.O. and Chua, K.E. (1969). Preliminary investigation of the exploitation of some potential nutritional resources by three sympatric tubificid oligochaetes. *Journal of The Fisheries Research Board of Canada* **26**, 2659-2668.

Brooker, M.P. and Edwards, R.W. (1974). Effects of the herbicide paraquat on the ecology of a reservoir III. Fauna and general discussion. *Freshwater Biology* **4**, 311-335.

Calow, P. (1974). Evidence for bacterial feeding in *Planorbis contortus* Linn. (Gastropoda : Pulmonata). *Proceedings of the*

Malacological Society of London **41**, 145-156.

Calow, P. (1977). Conversion efficiencies in heterotrophic organisms. *Biological Reviews* **52**, 385-409.

Carter, C.E. (1978). The fauna of the muddy sediments of Lough Neagh, with particular reference to eutrophication. *Freshwater Biology* **8**, 547-559.

Caspers, H. (1980). The relationship of saprobial conditions to massive populations of tubificids. In "Aquatic Oligochaete Biology" (Eds. R.O. Brinkhurst and D.G. Cook), pp.503-505. Plenum Press, New York.

Chatarpaul, L., Robinson, J.B. and Kaushik, N.K. (1979). Role of tubificid worms on nitrogen transformations in stream sediment. *Journal of the Fisheries Research Board of Canada* **36**, 673-678.

Chatarpaul, L., Robinson, J.B. and Kaushik, N.K. (1980). Effects of tubificid worms on denitrification and nitrification in stream sediment. *Canadian Journal of Fisheries and Aquatic Sciences* **37**, 656-663.

Chua, K.E. and Brinkhurst, R.O. (1973). Bacteria as potential nutritional resources for three sympatric species of tubificid oligochaetes. In "Estuarine Microbial Ecology" (Eds. L.H. Stevenson and R.R. Colwell), pp.513-517. University of South Carolina Press, Columbia.

Colwell, R.R. and Liston, J. (1962). The natural bacterial flora of certain marine invertebrates. *Journal of Insect Pathology* **4**, 23-33.

Coull, B.C. (1973). Estuarine Meiofauna: A review: Trophic relationships and microbial interactions. In "Estuarine Microbial Ecology" (Eds. L.H. Stevenson and R.R. Colwell), pp.499-512. University of South Carolina Press, Columbia.

Edwards, R.W. (1957). Vernal sloughing of sludge deposits in a sewage effluent channel. *Nature* **180**, 100.

Edwards, R.W. (1958). The effect of larvae of *Chironomus riparius* Meigen on the redox potentials of settled activated sludge. *The Annals of Applied Biology* **46**, 457-464.

Edwards, R.W. (1962). Some effects of plants and animals on the conditions in freshwater streams with particular reference to their oxygen balance. *International Journal of Air and Water Pollution* **6**, 505-520.

Edwards, R.W. and Rolley, H.L.J. (1965). Oxygen consumption of river muds. *Journal of Ecology* **53**, 1-19.

Edwards, R.W., Benson-Evans, K., Learner, M.A., Williams, P. and Williams, R. (1972). A biological survey of the River Taff. *Water Pollution Control* **71**, 144-166.

Fenchel, T. (1972). Aspects of decomposer food chains in marine benthos. *Verhandlungsbericht der Deutschen zoologischen gesellschaft* **65**, 14-22.

Fenchel, T. (1975). The quantitative importance of the benthic microfauna of an arctic tundra pond. *Hydrobiologia* **46**, 445-464.

Fenchel, T.M. (1978). The ecology of micro- and meiobenthos. *Annual Review of Ecology and Systematics* **9**, 99-121.

Fenchel, T. (1980). Suspension feeding in ciliated protozoa: feeding rates and their ecological significance. *Microbial Ecology* **6**, 13-25.

Fenchel, T.M. and Jørgensen, B.B. (1977). Detritus food chains of aquatic ecosystems: the role of bacteria. *Advances in Microbial Ecology* **1**. 1-58.

Fredeen, F.J.H. (1964). Bacteria as food for blackfly larvae (Diptera: Simuliidae) in laboratory cultures and in natural streams. *Canadian Journal of Zoology* **42**, 527-548.

Fry, J.C. and Staples, D.G. (1976). Distribution of *Bdellovibrio bacteriovorus* in sewage works, river water and sediments. *Applied and Environmental Microbiology* **31**, 469-474.

Fry, J.C., Brooker, M.P. and Thomas, P.L. (1973). Changes in the microbial populations of a reservoir treated with the herbicide paraquat. *Water Research* **7**, 395-407.

Gerlach, S.A. (1978). Food-chain relationships in subtidal silty sand marine sediments and the role of meiofauna in stimulating bacterial productivity. *Oceologia* (Berlin) **33**, 55-69.

Granéli, W. (1979). The influence of *Chironomus plumosus* larvae on the oxygen uptake of sediment. *Archiv für Hydrobiologie* **87**, 385-403.

Hargrave, B.T. (1976). The central role of invertebrate faeces in sediment decomposition. In "The role of Terrestrial and Aquatic Organisms in Decomposition Processes" (Eds. J.M. Anderson and A. Macfadyen), pp.301-321. Blackwell Scientific Publications, Oxford.

Harper, R.M., Fry, J.C. and Learner, M.A. (1981*a*). A bacteriological investigation to elucidate the feeding biology of *Nais variabilis* (Oligochaeta: Naididae). *Freshwater Biology* **11**, 227-236.

Harper, R.M., Fry, J.C. and Learner, M.A. (1981*b*). Digestion of bacteria by *Nais variabilis* (oligochaeta) as established by autoradiography. *Oikos* **36**, 211-218.

Henriksen, K., Hansen, J.I. and Blackburn, T.H. (1980). The influence of benthic infauna on exchange rates of inorganic nitrogen between sediment and water. *Ophelia Supplement* **1**, 249-256.

Hylleberg, J. and Henriksen, K. (1980). The central role of bioturbation in sediment mineralization and element recycling. *Ophelia Supplement* **1**, 1-16.

Jacobsen, O.S. (1977). The influence of bottom fauna density on the exchange rates of phosphate and inorganic nitrogen in a eutrophic profundal sediment. *Proceedings of the Nordic Sediment Symposium* **5**, 39-49.

Johnson, M.G. and Brinkhurst, R.O. (1971). Production of benthic macroinvertebrates of Bay of Quinte and Lake Ontario. *Journal of the Fisheries Research Board of Canada*. **28**, 1699-1714.

Jones, J.G. (1976). The microbiology and decomposition of seston in open water and experimental enclosures in a productive lake. *Journal of Ecology* **64**, 241-278.

Jones, J.G. (1980). Some differences in the microbiology of profundal and littoral lake sediments. *Journal of General Microbiology* **117**, 285-292.

Jones, J.G., Orlandi, M.J.L.G. and Simon, B.M. (1979). A microbiological study of sediments from the Cumbrian Lakes. *Journal of General Microbiology* **115**, 37-48.

Jørgensen, C.B. (1976). August Pütter, August Krogh and modern ideas on the use of dissolved organic matter in aquatic environments. *Biological Reviews* **51**, 291-328.

Jørgensen, B.B. (1980). Mineralization and the bacterial cycling of carbon, nitrogen and sulphur in marine sediments. In "Contemporary Microbial Ecology" (Eds. D.C. Elwood, J.N. Hedger, M.J. Latham,

J.M. Lynch and J.H. Slater), pp.239-251. Academic Press, London.
Koehn, T. and Frank, C. (1980). Effect of thermal pollution on the chironomic fauna in an urban channel. In "Chironomidae-Ecology, Systematics, Cytology and Physiology" (Ed. D.A. Murray), pp.187-194. Pergamon Press, Oxford.
Krezoski, J.R., Mozley, S.C. and Robbins, J.A. (1978). Influence of benthic macroinvertebrates on mixing of profundal sediments in south eastern Lake Huron. *Limnology and Oceanography* **23**, 1011-1016.
Learner, M.A., Lochhead, G. and Hughes, B.D. (1978). A review of the biology of British Naididae (Oligochaeta) with emphasis on the lotic environment. *Freshwater Biology* **8**, 357-375.
Lopez, G.R., Levinton, J.S. and Slobodkin, L.B. (1977). The effect of grazing by detritivore *Orchestia grillus* on *Spartina* litter and its associated microbial community. *Oecologia* (Berlin) **30**, 111-127.
Lopez, G., Riemann, F. and Schrage, M. (1979). Feeding biology of the brackish-water oncholaimid nematode *Adoncholaimus thalassophygas*. *Marine Biology* **54**, 311-318.
Marcus, J.H., Sutcliffe, D.W. and Willoughby, L.G. (1978). Feeding and growth of *Asellus aquaticus* (Isopoda) on food items from the littoral of Windermere, including green leaves of *Elodea canadensis*. *Freshwater Biology* **8**, 505-519.
Marshall, K.C. (1980). Reactions of microorganisms, ions and macromolecules at interfaces. In "Contemporary Microbial Ecology" (Eds. D.C. Ellwood, J.N. Hedger, M.J. Latham, J.M. Lynch and J.H. Slater), pp.93-106. Academic Press, London.
McCall, P.L. and Fisher, J.B. (1980). Effects of tubificid oligochaetes on physical and chemical properties of Lake Erie sediments. In "Aquatic Oligochaete Biology" (Eds. R.O. Brinkhurst and D.G. Cook), pp.253-317.
McLachlan, A.J. and Dickinson, C.H. (1977). Microorganisms as a factor in the distribution of *Chironomus lugubris* Zetterstedt in a bog lake. *Archiv für Hydrobiologie* **80**, 133-146.
Monakov, A.V. (1972). Review of studies on feeding of aquatic invertebrates conducted at the Institute of Biology of Inland Waters, Academy of Science, USSR. *Journal of the Fisheries Research Board of Canada* **29**, 363-383.
Newell, R. (1965). The role of detritus in the nutrition of two marine deposit feeders, the prosobranch *Hydrobia ulvae* and the bivalve *Macoma balthica*. *Proceedings of the Zoological Society of London* **144**, 25-45.
Palmer, M.F. (1968). Aspects of the respiratory physiology of *Tubifex tubifex* in relation to its ecology. *Journal of Zoology, London* **154**, 463-473.
Pearson, T.H. and Rosenberg, R. (1978). Macrobenthic succession in relation to organic enrichment and pollution of the marine environment. *Oceanography and Marine Biology an Annual Review* **16**, 229-311.
Pomeroy, L.R. (1979). Microbial roles in aquatic food webs. In "Aquatic Microbial Ecology" (Eds. R.R. Colwell and J. Foster), pp.85-109. University of Maryland, College Park.
Poole, N.J. and Wildish, D.J. (1979). Polysaccharide degradation in estuaries. In "Microbial Polysaccharides and Polysaccharases" (Eds. R.C.W. Berkeley, G.W. Gooday and D.C. Ellwood), pp.399-416. Academic Press, London.
Potter, D.W.B. and Learner, M.A. (1974). A study of the benthic

macro-invertebrates of a shallow eutrophic reservoir in South Wales with emphasis on Chironomidae (Diptera); their life-histories and production. *Archiv für Hydrobiologie* **74**, 186-226.

Reish, D.J. (1979). Bristle worms (Annelida : Polychaeta). In "Pollution Ecology of Estuarine Invertebrates" (Eds. C.W. Hart and S.L.H. Fuller), pp.77-125. Academic Press, London.

Revsbech, N.P., Sørensen, J. and Blackburn, T.H. (1980). Distribution of oxygen in marine sediments measured with microelectrodes. *Limnology and Oceanography* **25**, 403-411.

Riemann, F. and Schrage, M. (1978). The mucus-trap hypothesis on feeding of aquatic nematodes and implications for biodegradation and sediment texture. *Oecologia* (Berlin) **34**, 75-88.

Ripley, M.P. (1980). The relation of dry weight and temperature to respiration in some benthic chironomic species in Lough Neagh. In "Chironomidae-Ecology, Systematics, Cytology and Physiology" (Ed. D.A. Murray), pp.81-88. Pergamon Press, Oxford.

Sochard, M.R., Wilson, D.F., Austin, B. and Colwell, R.R. (1979). Bacteria associated with the surface and gut of marine copepods. *Applied and Environmental Microbiology* **37**, 750-759.

Sørensen, J., Jørgensen, B.B. and Revsbech, N.P. (1979). A comparison of oxygen, nitrate and sulphate respiration in coastal marine sediments. *Microbial Ecology* **5**, 105-115.

Tietjen, J.H. (1979). Microbial-meiofaunal inter-relationships: A review. In "Aquatic Microbial Ecology" (Eds. R.R. Colwell and J. Foster), pp.130-140. University of Maryland, College Park.

Titmus, G. and Badcock, R.M. (1980). Production and emergence of chironomids in a wet gravel pit. In "Chironomidae-Ecology, Systematics, Cytology and Physiology" (Ed. D.A. Murray), pp.299-305. Pergamon Press, Oxford.

Toscano, R.J. and McLachlan, A.J. (1980). Chironomids and particles: Microorganisms and chironomid distribution in a peaty upland river. In "Chironomidae-Ecology, Systematics, Cytology and Physiology" (Ed. D.A. Murray), pp.171-177. Pergamon Press, Oxford.

Walshe, B.M. (1951). The feeding habits of certain chironomid larvae (subfamily Tendipedinae). *Proceedings of the Zoological Society of London* **121**, 63-79.

Ward, G.M. and Cummins, K.W. (1979). Effects of food quality on growth of a stream detritivore, *Paratendipes albimanus* (Meigen) (Diptera : Chironomidae). *Ecology* **60**, 57-64.

Waters, T.F. (1977). Secondary production in inland waters. *Advances in Ecological Research* **10**, 91-164.

Wavre, M. and Brinkhurst, R.O. (1971). Interactions between some tubificid oligochaetes and bacteria found in the sediments of Toronto Harbour, Ontario. *Journal of the Fisheries Research Board of Canada* **28**, 335-341.

Wetzel, R.G. (1975). "Limnology". W.B. Saunders, Philadelphia.

Whitten, B.K. and Goodnight, C.J. (1966). The comparative chemical composition of two aquatic oligochaetes. *Comparative Biochemistry and Physiology* **17**, 1205-1207.

Whitten, B.K. and Goodnight, C.J. (1969). The role of tubificid worms in the transfer of radioactive phosphorus in an aquatic ecosystem. In "Symposium on Radioecology - Proceedings of Second National Symposium" (Eds. D.J. Nelson and F.C. Evans), pp.270-277. U.S. Atomic Energy Commission, Ann Arbor.

Willoughby, L.G. and Marcus, J.H. (1979). Feeding and growth of the isopod *Asellus aquaticus* on actinomycetes, considered as model filamentous bacteria. *Freshwater Biology* **9**, 441-449.

Wood, L.W. (1975). Role of oligochaetes in the circulation of water and solutes across the mud-water interface. *Verhandlung der Internationalen Vereinigung für theoretische und angewandte Limnologies* **19**, 1530-1533.

Chapter 8

EARLY TRANSFORMATIONS OF ISOPRENOID COMPOUNDS IN SURFACE SEDIMENTARY ENVIRONMENTS

J.R. MAXWELL and A.M.K. WARDROPER

Organic Geochemistry Unit, University of Bristol, School of Chemistry, Bristol, UK

Introduction

One of the most remarkable features about the distributions of many of the lipid classes in recent sediments is their basic similarity. This relates in part to the fact that any aquatic sediment acts as a sink for the organic matter from a variety of organisms. It also relates to the presence of bacteria which are active both in the sediment itself and in the overlying water column, and which (i) contribute a variety of their own characteristic natural products (ii) mineralize certain organic compounds contributed by other organisms (iii) modify the structures (for example, reduction of carbon-carbon double bonds) of other lipids from these organisms. These processes are perhaps most simply illustrated by the fatty acid distributions of surface sediments. A general feature here is the rapid decrease in the abundance of unsaturated fatty acids relative to their saturated counterparts with increasing depth in the sediment [Rhead *et al.*, 1971, 1972; Farrington and Quinn, 1973; Johnson and Calder, 1973; Johns *et al.*, 1978]. *In situ* incubation of Z-9, 10-^{3}H, 1-^{14}C-octadec-9-enoic acid (oleic acid) in sediment from the Severn Estuary [Rhead *et al.*, 1971, 1972; Gaskell *et al.*, 1976] revealed the short term fate of this acid (Fig. 1) and therefore provided information about the origins of the fatty acid distribution in this sediment. The changes are summarized as follows: (i) reduction of the Δ^9-double bond (ii) degradation by the well known β-oxidation pathway to lower molecular weight acids (for example, $C_{14:0}$) (iii) resynthesis of saturated straight chain and branched fatty acids from acetyl CoA (iv) incorporation into cell constituents or loss as CO_2. The operation of such microbial transformations and processes appears, therefore, to explain the fatty acid distributions in many recent sediments. For example, those of both fatty acids and hydroxy

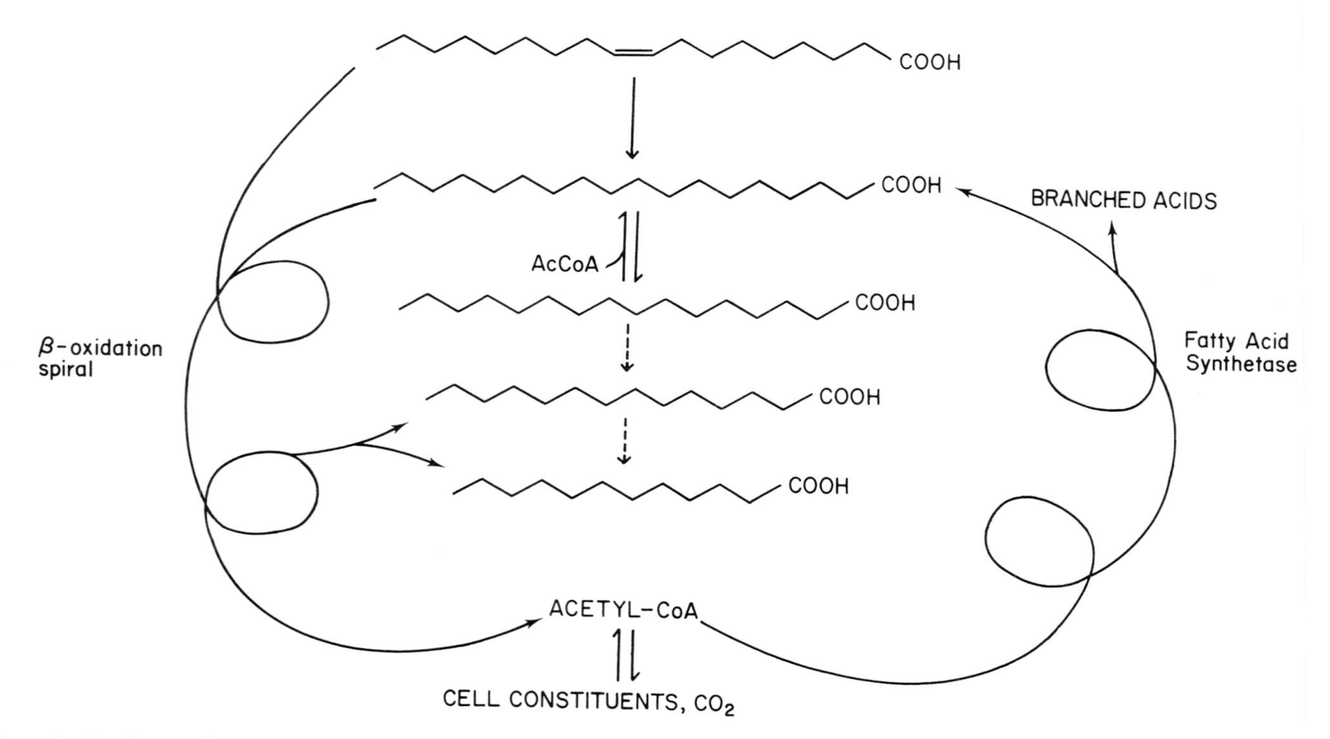

Fig. 1 Outline of pathways for oleic acid metabolism in Severn Estuary sediment [adapted from Rhead *et al.*, 1972].

fatty acids at different depths in a core of diatomaceous ooze from Walvis Bay, off S.W. Africa, bore little resemblance to those observed in diatoms but were explicable in terms of an origin from bacterial reworking of the algal input [Boon *et al.*, 1975, 1977, 1978].

Recent developments in high resolution gas chromatography and the resulting improvements in the separation of structural and stereochemical isomers emphasize further that sedimentary fatty acid distributions are the net result of inputs from a diversity of organisms and from bacterial degradation and synthesis [for example, Perry *et al.*, 1979].

In this paper, discussion is restricted, however, to lipids biosynthesised *via* the mevalonate (isoprenoid) pathway rather than those biosynthesised by the acetate pathway. The latter compounds have, in general, a lower structural specificity (both gross carbon skeleton and stereochemistry) and therefore lower "information content" in terms of organic geochemical studies of sedimentary organic matter. One fundamental aspect of organic geochemical studies is to attempt to interpret lipid distributions of recent and ancient sediments in terms of the inputs of organic matter from different contributing organisms and therefore of the environment at the time of deposition. For example, the occurrence of 4,23,24-trimethylcholest-22-ene (Ia) as a relatively abundant alkene in a Miocene claystone of *ca.* 20 million years old has been taken as indicative of a dinoflagellate input [McEvoy *et al.*, 1981]. This compound is not biosynthesised by organisms and is likely to be derived from the corresponding sterol (Ib) which is believed to be unique to dinoflagellates [Boon *et al.*, 1979]. It is therefore essential to understand the biochemical and chemical pathways which alter the precursor natural products in the sedimentary column over geological time.

I

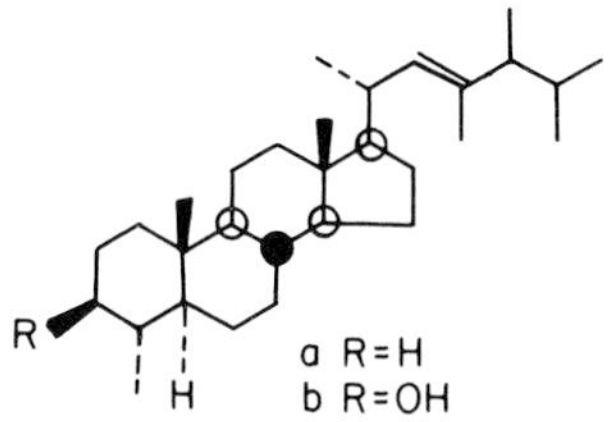

a R=H
b R=OH

II

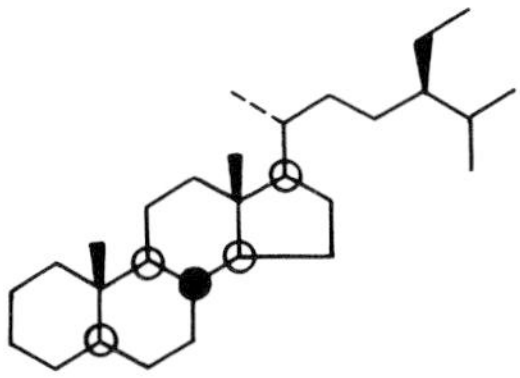

These pathways encompass the time span from the incorporation of the natural product into the sediment to the stage at which the derived products (biological marker compounds) can still be recognized, and their structures correlated with the original biosynthesised compounds. For example, tetracyclic alkanes (such as II) derived from sterol precursors occur almost ubiquitously as complex mixtures in sedimentary rocks and petroleums [for example, Mulheirn and Ryback, 1975]. The pathway linking

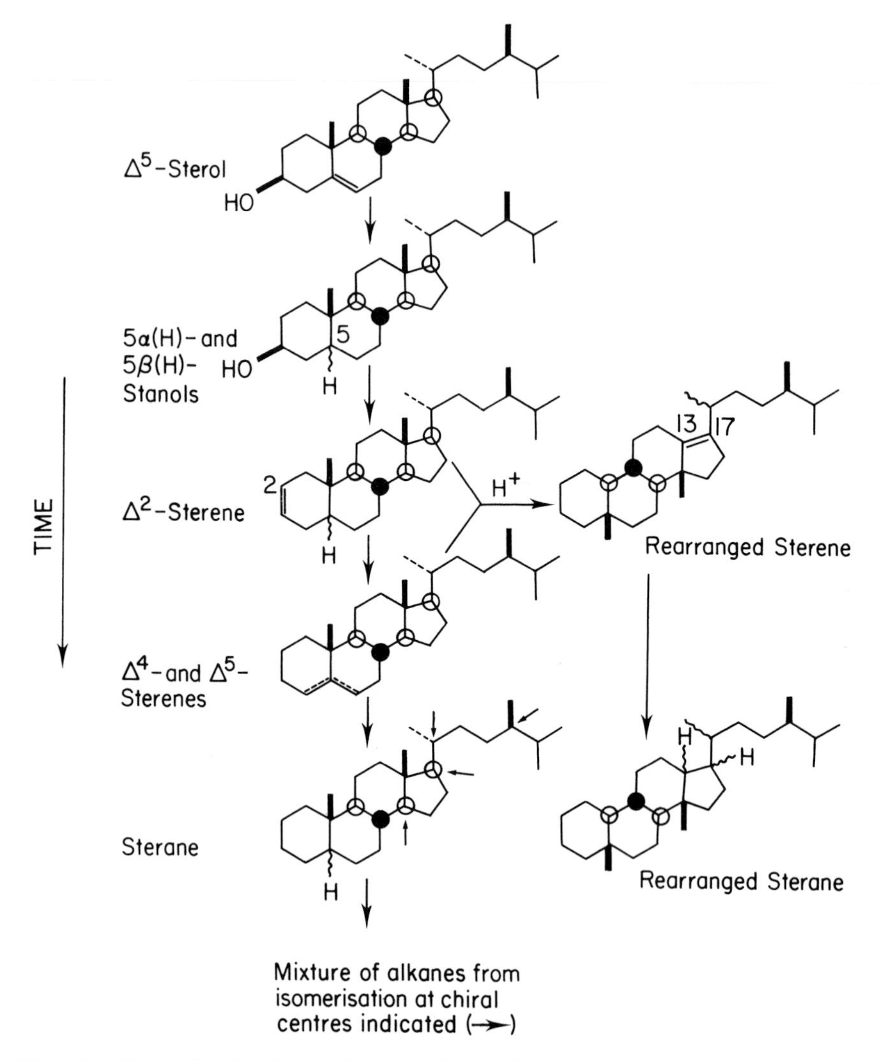

Fig. 2 Simplified scheme for the fate of sterols in sediments over geological time.

sterols to the complex mixtures of steroidal alkanes in sedimentary rocks and petroleums is shown in a simplified form in Fig. 2. The early stage intermediates shown have been found in surface sediments, and include the saturated alcohols [stanols; Gaskell and Eglinton, 1976; Nishimura and Koyama, 1976], Δ^2-sterenes [Dastillung and Albrecht, 1977; Gagosian and Farrington, 1978; Quirk *et al.*, 1980*b*] and the $\Delta^{13(17)}$-rearranged sterenes [Dastillung and Albrecht, 1977]. The complexity of the resulting alkanes arises from (i) the structural diversity of the original sterols contributed to the sediment [Wardroper *et al.*, 1978; Lee *et al.*, 1979] (ii) a late stage (clay catalysed?) isomerization at a number of chiral centres in the first formed alkanes which have, until this stage, retained the original stereochemistry of the sterols [examples include Mulheirn and Ryback, 1975; Mackenzie *et al.*, 1980] (iii) the presence of configurational isomers of alkanes with a rearranged skeleton resulting from reduction of the corresponding rearranged alkenes formed at an earlier stage [Ensminger *et al.*, 1978]. Studies of the distributions of these alkanes not only allows an understanding of the sedimentary fate of sterols but in a wider context have application to (i) the correlation of petroleums with each other and with their source rocks in oil exploration [for example, Seifert and Moldowan, 1978] (ii) the identification of oil spillages in the environment [Albaiges, 1980] (iii) assessment of the thermal history of sedimentary rocks and petroleums [Seifert and Moldowan, 1978; Mackenzie *et al.*, 1980]. A key to understanding fully this and other sedimentary pathways is the elucidation of the early stages by way of identifying the sequence and mechanisms of the reactions and the agents (chemical and biological) responsible. At present, three approaches have been used by chemists:

1) Comparison of the lipid distributions of surface sediments with those of organisms in general or organisms known or suspected to contribute their organic matter to the sediment. For example, Δ^2-sterenes occur widely in surface marine sediments [Fig. 2; Dastillung and Albrecht, 1977; Gagosian and Farrington, 1978] but they have never been found in organisms. By default, they are therefore assumed to arise from transformations occurring in these sediments. The mechanism is unknown.

2) "Bacterial decay" experiments with specific organisms whereby an organism, such as an alga, known to contribute to the sediment, is allowed to incubate either *in situ* [Quirk *et al.*, 1980*a*] or in the laboratory under conditions thought to approximate to those in the sediment [Cranwell, 1976, 1979]. Changes in lipid distributions before and after incubation then allow possible transformations in the sediment under natural conditions to be inferred.

3) Incubation *in situ* or in the laboratory of precursor compounds, preferably radiolabelled, and identification of the products. For example, one of the first sedimentary transformations of sterols (Fig. 2) is microbial reduction (see below) of the Δ^5-double bond to give the equivalent stanols, as shown by the conversion of the 4-^{14}C-cholest-5-en-3β-ol (III) to 5α(H)- and 5β(H)-cholestan-3β-ol (IV) in a lake sediment [Gaskell and Eglinton, 1975].

Other isoprenoid compounds found in sedimentary rocks and petroleums include (i) pentacyclic triterpenoid alkanes (for example, V), thought to be derived from functionalized precursors in bacteria present at the time

III

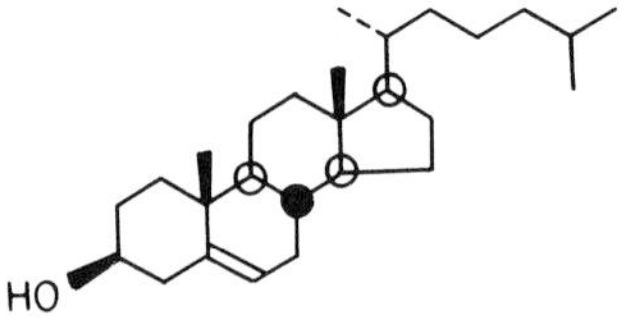

IV

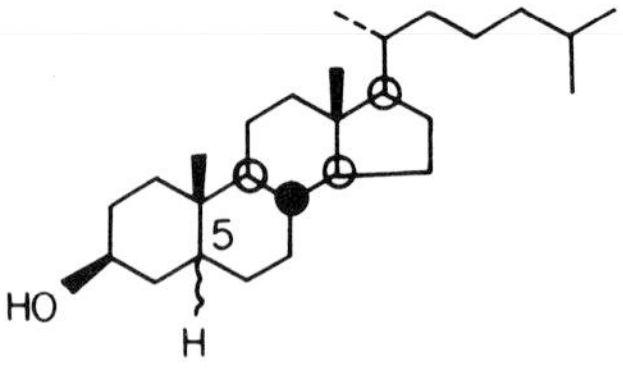

of deposition [Van Dorsselaer *et al.*, 1974; Ourisson *et al.*, 1979], (ii) acyclic isoprenoids (for example, VI) which arise in part from the phytol side chain of various chlorophylls [Maxwell *et al.*, 1973].

V

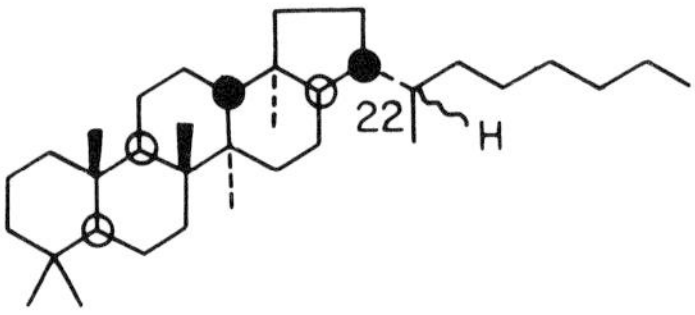

VI

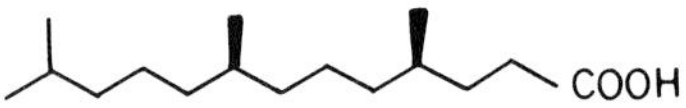

In the following sections some early transformations within the steroid, triterpenoid and acyclic isoprenoid series are discussed. Emphasis is placed on the methods used and on the limited information available at present about the mechanisms which are involved.

Steroids

Figure 3 summarizes the transformations of Δ^5-sterols which are thought to occur in surface sediments and is based mainly on reported occurrences of the various compound types in such sediments. The scheme is a generalized one and does not take account of differences in depositional environments. For example, Δ^2-sterenes have not been observed or are present only in trace amounts in oxic marine sediments or in lake sediments, whereas they are abundant in anoxic marine sediments [Gagosian *et al.*, 1980; Quirk *et al.*, 1980*b*]. Although in most cases the mechanisms are not clear, the overall sedimentary fate of sterols is probably the best understood in comparison with other lipid classes.

The first transformation to be recognized (and possibly the first to occur) is reduction of the Δ^5-double bond. This was inferred from the increase in the 5α(H)-stanol/Δ^5-sterol ratio with increasing sediment depth and the parallelism in carbon skeletal distribution between the two classes in lacustrine sediments [Ogura and Hanya, 1973; Gaskell and Eglinton, 1976; Nishimura and Koyama, 1976]. Direct evidence for the reduction was obtained by incubation, in a eutrophic lake sediment, of radiolabelled cholest-5-en-3β-ol (cholesterol, III) in which 5α(H)- and 5β(H)-cholestan-3β-ol (IV) were formed in low yield

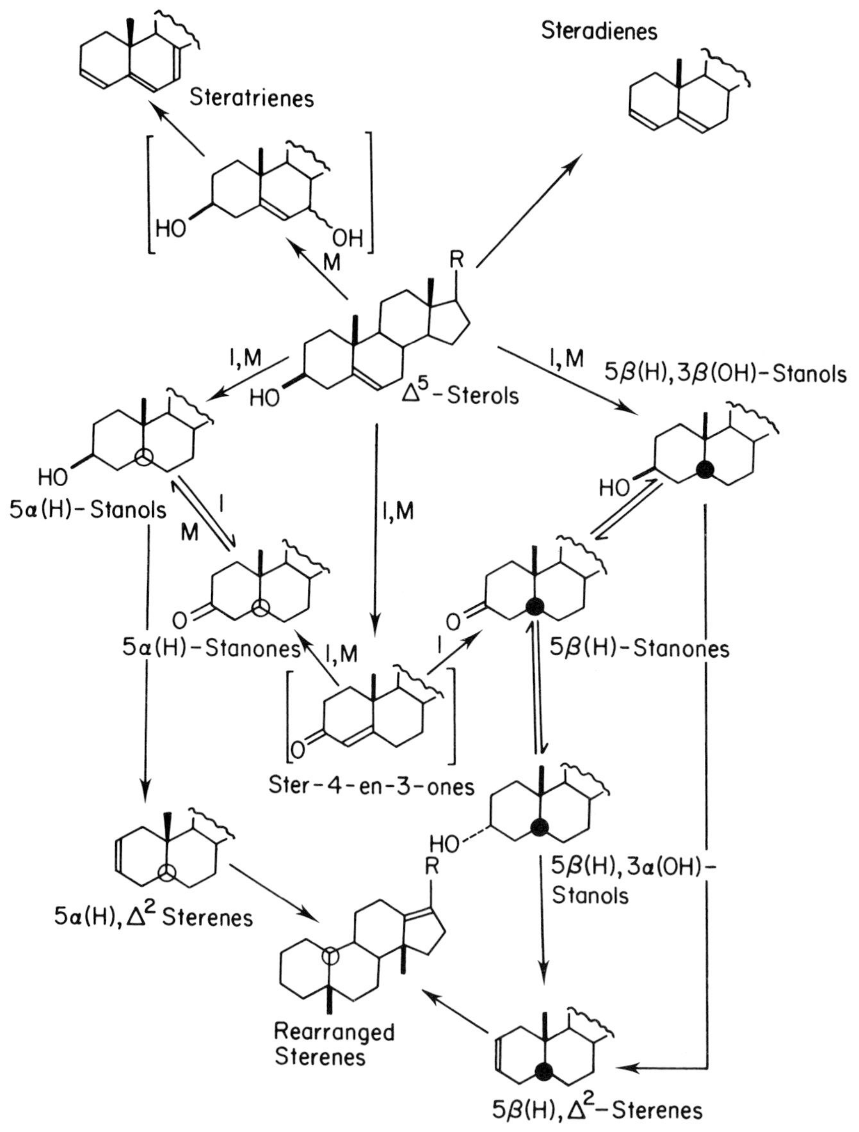

Fig. 3 Proposed scheme for the transformations of Δ^5-sterols in recent sediments [adapted from Gagosian *et al.*, 1980, and Quirk *et al.*, 1980*b*]. Compounds in brackets have not been isolated from surface sediments. I - transformation observed in radiolabelled incubation studies in sedimentary environments. M - transformations observed in microbial cultures [Björkhem and Gustafsson, 1971; Kramli and Horvath, 1948]. R = side chain.

TABLE 1

Incubation of 4-^{14}C-cholest-5-en-3β-ol in Rostherne sediment and anaerobic sewage sludge [from Gaskell and Eglinton, 1975]

Radiolabel	Experiment*		
	A (sediment)	B (sediment)	C (sewage)
Activity injected (μCi)	14.67	3.10	3.61
Incubation products (%):			
Total	69	29	77
4-^{14}C-cholest-5-en-3β-ol	45	n.d.+	28
4-^{14}C-5α(H)-cholestan-3β-ol	0.47	0.09	0.66
4-^{14}C-5β(H)-cholestan-3β-ol	0.11	0.03	1.2

* A, incubation (90 days, 10°C) in intact core in the laboratory (dark);
B, incubation (65 days, 5.4°C) at lake bottom, injection at 4 cm sediment depth;
C, incubation (28 days, 37°C) in the laboratory (dark).

\+ Not determined.

[Gaskell and Eglinton, 1975]. The results of the radiolabelled experiments are summarized in Table 1. The 5α(H)-isomer predominated in the sediment and as a radiolabelled product, whereas in anaerobic sewage sludge the 5β(H)-isomer was the more abundant product. This evidence indicates that the reduction of the Δ^5-double bond is a microbial, probably bacterial, process. Such a mechanism has parallels in biological systems and has been observed in certain bacterial [Schubert and Kaufmann, 1965; Björkhem and Gustafsson, 1971; Rosenfeld and Hellman, 1971; Eyssen *et al.*, 1973; Eyssen and Parmentier, 1974] and other systems [Smith *et al.*, 1972; Stohs and Haggerty, 1973]. Comparison of the results of the sediment and sewage incubations suggest that the 5α(H)-/5β(H)-stanol ratio is dependent on the particular microbial population. It is known that faecal bacteria in higher animals produce 5β(H)-cholestan-3β-ol from cholesterol in preference to the 5α(H)-isomer [Rosenfeld and Gallagher, 1964]. The sediment incubation studies highlight a general problem in this area. Radiolabelling studies can demonstrate that a particular transformation occurs in a sediment but in no case have any bacteriological studies been carried out in parallel to the chemical studies to identify the organisms responsible.

The appearance of stanols in surface sediments has also been investigated in "decay" experiments [Quirk *et al.*, 1980*b*]. The following samples were examined in relation to a study of a eutrophic lake, Rostherne Mere (Cheshire) - (i) a sample (collected from the lake surface) comprising mainly the blue-green alga *Microcystis aeruginosa* (known to be a major contributor of organic matter to the sediment), with minor amounts of other algae, (ii) a similar sample allowed to decay on the lake surface, (iii) a similar sample allowed to decay as in (ii) and then on the lake bottom, (iv) the surface sediment (0 to 20 cm). The "decay" experiments were intended to represent the natural life of the alga, in that it lives on the lake surface during summer and sinks to the bottom during winter. Figure 4 shows the distribution of the major monounsaturated sterols with a Δ^5-double bond and their 5α(H)- and 5β(H)-counterparts. A general decrease with "decay" in the 5α(H)-stanols and in the 5β(H)-stanols (3β-OH isomer; cf. Fig. 3) relative to the corresponding Δ^5-sterols is apparent. In the sediment itself, both the 3α- and 3β-OH isomers of the 5β(H)-stanols were present. In a predominantly sphagnum peat [Lyne of Skene, Aberdeenshire; Quirk *et al.*, 1980*b*] a similar sequence was observed; the major contributing organism, *Sphagnum cuspidatum* contains no 5β(H)-stanols, the moss base (the previous year's moss "decayed" naturally for *ca.* 1 year) contained low concentrations (< 2 ppm, 3β-OH isomer only) and the peat itself (40 - 50 cm) had both the 3β-OH and 3α-OH alcohols (*ca.* 70 ppm).

VII

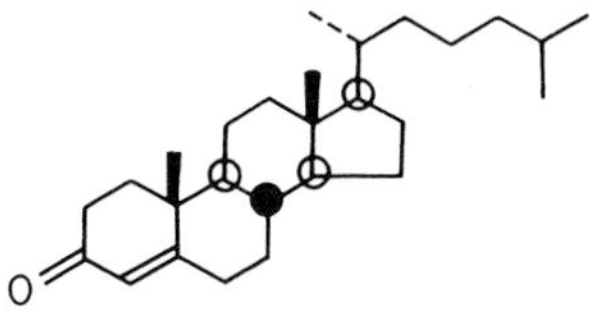

VIII

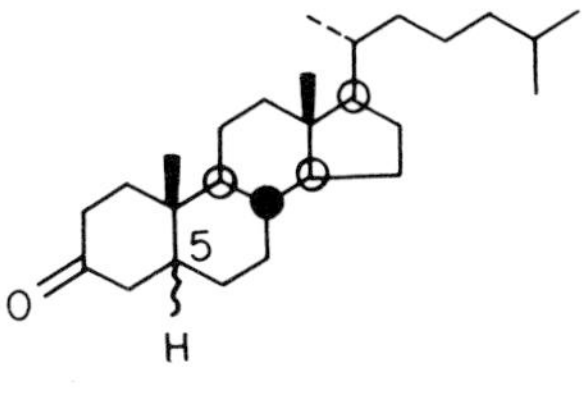

These and the radiolabelling studies suggest that a complex oxidation-reduction system operates on the steroidal alcohols in recent sediments, with saturated and unsaturated ketones as possible intermediates. For example, a plausible pathway to the 5β(H), 3α(OH)-stanols which have never been found in any organism, is *via* the 5β(H)-stanones (Fig. 3). In the sediment radiolabelling studies involving cholesterol (III), other products which were tentatively identified included cholest-4-en-3-one (VII) and 5α(H)- and 5β(H)-cholestan-3-one (VIII). Therefore, unsaturated ketones are likely to be intermediates on the pathway to the saturated alcohols in sediments (Fig. 3), although they have not yet been reported in surface sediments. It has been suggested that this may result from their low steady-state concentration in the

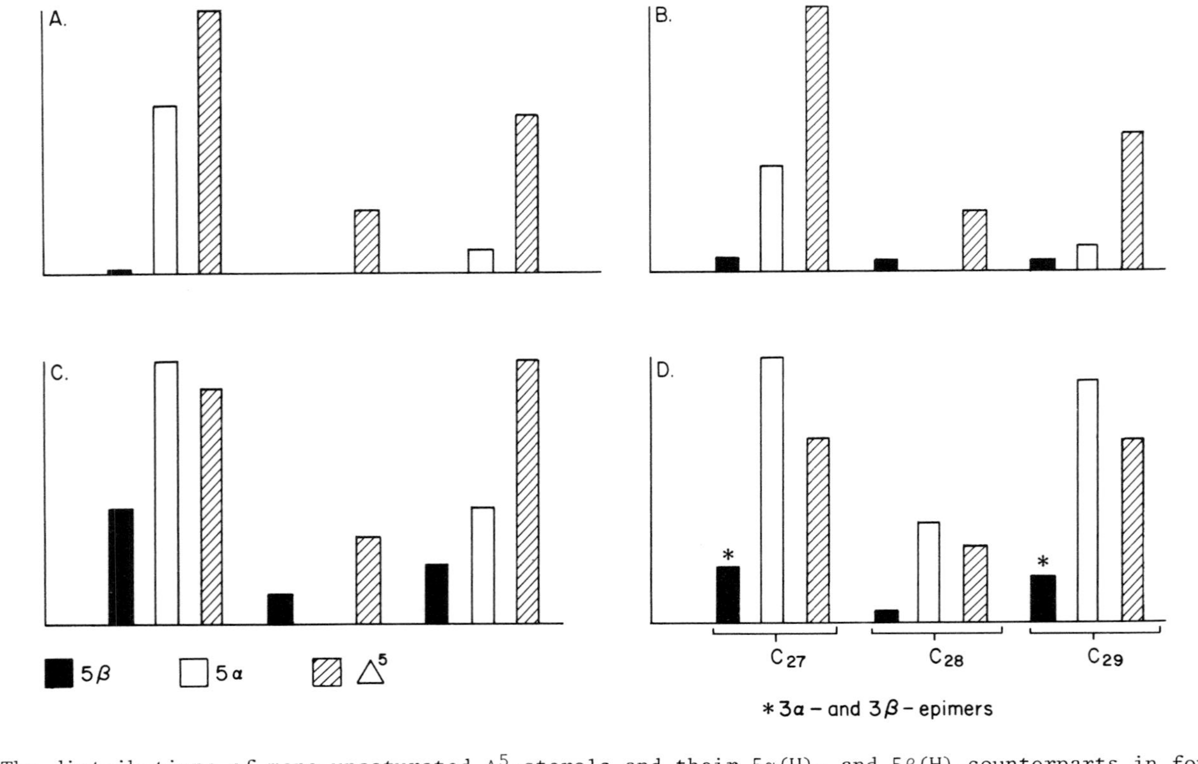

Fig. 4 The distributions of mono-unsaturated Δ^5-sterols and their 5α(H)- and 5β(H)-counterparts in four samples: **A.** *M. aeruginosa* collected from Rostherne Mere; **B.** *M. aeruginosa* decayed for 5 months on the lake surface (1 m water depth); **C.** *M. aeruginosa* decayed for 5 months on the lake surface and 1 year on the lake bottom; **D.** Rostherne Mere surface sediment (0 to 20 cm).

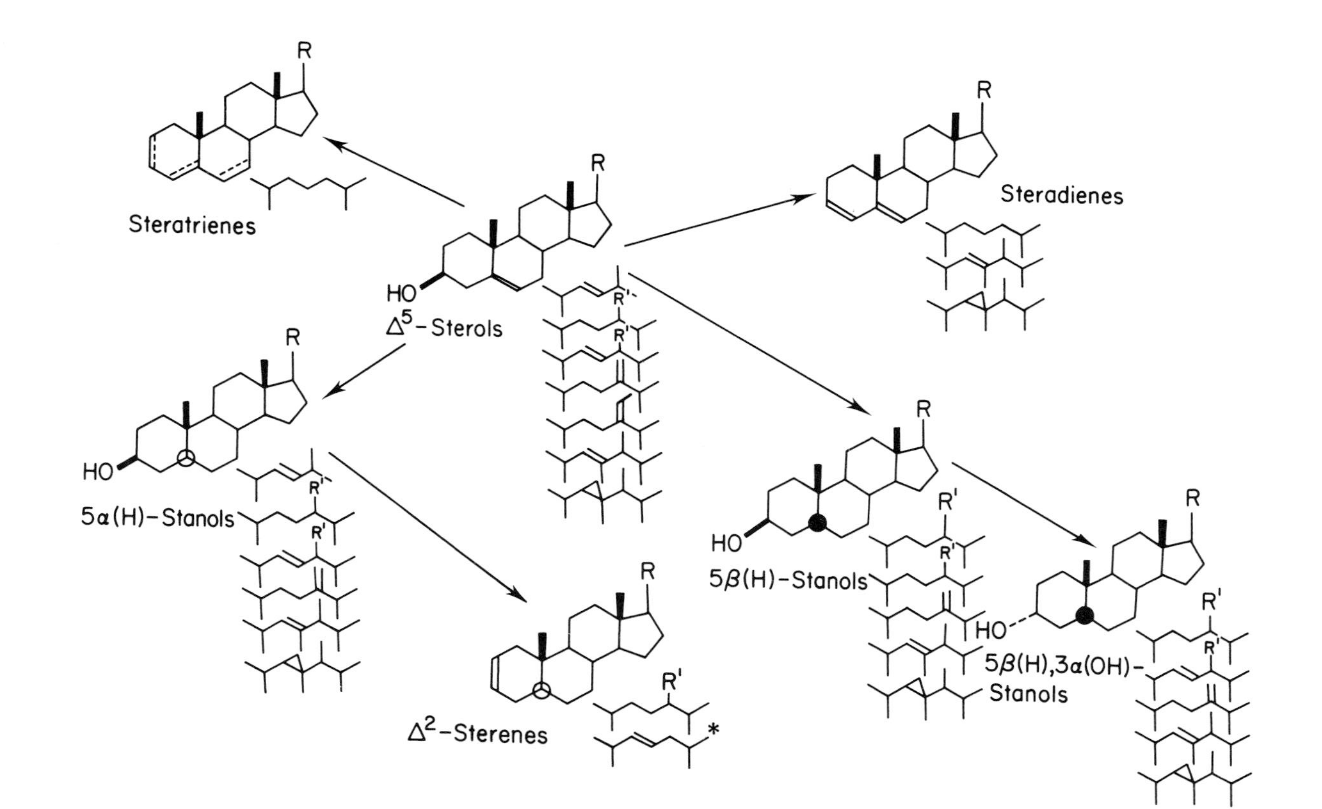

Fig. 5 Sterol transformations in a marine surface sediment (Walvis Bay, Namibia) inferred from the distributions of steroidal derivatives present. R = side chain, $R^1 = H, CH_3, C_2H_5$, * position of nuclear double bond unknown.

reaction sequence [Gagosian *et al.*, 1980]. The intermediacy of ketones in the conversion of Δ^5-sterols to the corresponding stanols (5α(H) and 5β(H)) has been demonstrated in various organisms, including bacteria [Björkhem and Gustafsson, 1971; Stohs and Haggerty, 1973; Smith *et al.*, 1972; Eyssen *et al.*, 1973]. In sediment studies, incubations with the proposed intermediates in Fig. 3 remain, however, to be carried out, although *in situ* incubations of 4-^{14}C-5α(H)-cholestan-3β-ol (IV) in an algal mat showed the formation of 5α(H)-cholestan-3-one [VIII; Edmunds *et al.*, 1980]. These results suggest that stanones and stanols in sediments are interconvertible by an oxidation-reduction mechanism (Fig. 3). It is unlikely that the saturated ketones are contributed by organisms directly to sediments as reports of such ketones in organisms are rare [Withers *et al.*, 1978].

Figure 5 summarizes the sterol transformations in a diatomaceous ooze from Walvis Bay, inferred in this case solely from the distributions of the different compound classes, and shows the particular side chains which were identified [Wardroper *et al.*, 1978; Quirk *et al.*, 1980*b*]. Virtually every sterol with a Δ^5-double bond was accompanied by the equivalent 5α(H)- and 5β(H)-stanols, with every 5β(H)-compound having the 3β-OH configuration being accompanied, in slightly higher abundance, by the more stable isomer with the 3α-OH configuration. These results are in accord with the incubation and "decay" results shown above. The Walvis Bay distributions (Fig. 5) also serve to illustrate the formation of the Δ^2-sterenes, steradienes and steratrienes, which have been previously reported [Dastillung and Albrecht, 1977; Gagosian and Farrington, 1978]. The difference in the side chain structures present in the Δ^2-sterene, $\Delta^{3,5}$-diene and triene classes shows a high degree of selectivity in the formation of the hydrocarbons from the original sterols. This selectivity suggests a microbial involvement rather than a chemical process in the sediment. The steradienes are presumed to be produced by direct dehydration of the Δ^5-sterols, whereas the Δ^2-sterenes arise from dehydration of the 5α(H)-stanols. The cholestatriene (Fig. 3, Fig. 5), whose structure remains to be proved, has been proposed as being formed by dehydration of a C_{27} sterol with a diunsaturated ring system, or by hydroxylation of a Δ^5-sterol and subsequent dehydration [Fig. 3; Gagosian *et al.*, 1980]. Microbial hydroxylation of the steroidal nucleus has been reported [Iizuka and Naito, 1967; Holland and Diakow, 1979]. Neither the hydroxylated intermediate nor the sterol with a diunsaturated nucleus have been observed in sediments where the cholestatriene has been found.

It has been noted that in certain recent marine sediments, rearranged sterenes have been formed from monoolefinic sterenes [Figs. 2 and 3; Dastillung and Albrecht, 1977]. This is an example of a steroidal transformation that has a direct chemical analogy, and can be achieved in the laboratory under acid conditions, or by heating with a clay catalyst [Rubinstein *et al.*, 1975]. The sediment transformation is therefore assumed to be chemical rather than biological. In general it must be emphasized, however, that where circumstantial evidence indicates microbial involvement in most of the early sedimentary sterol transformations, the organisms responsible have never been isolated or identified.

Triterpenoids

In terms of geochemical occurrence the most important triterpenoids by far are those of the hopane family (for example, V). Like the steranes, the distributions of hopanoid alkanes in sedimentary rocks and petroleums have application in petroleum exploration and in pollution studies because of their high structural specificity. To date, over a hundred compounds *in toto* of, or closely related to, the hopane skeleton, have been identified in sedimentary organic matter of various origins and ages. Members of the series are therefore found as ubiquitous and often abundant components of all sedimentary environments, both recent and ancient, and are thought to be derived primarily from hopanoid precursors present originally in bacteria, and to a lesser extent in blue-green algae [Ourisson *et al.*, 1979]. Indeed, hopanoid derivatives as a class are one of the most abundant types of organic compound on the Earth's surface. They are found as a variety of compound classes, including alkanes (for example, V), alkenes (for example, IX), aromatic hydrocarbons (for example, XV), aldehydes (for example, Xa), ketones (for example, XI), alcohols (for example, Xb) and acids (for example, Xc) [Van Dorsselaer *et al.*, 1974; Ourisson *et al.*, 1979; Dastillung *et al.*, 1980*a*,*b*]. In only a few cases, such as hop-22(29)-ene (IX), are the components present in surface sediments found also in organisms, where they represent a direct contribution. Most of the

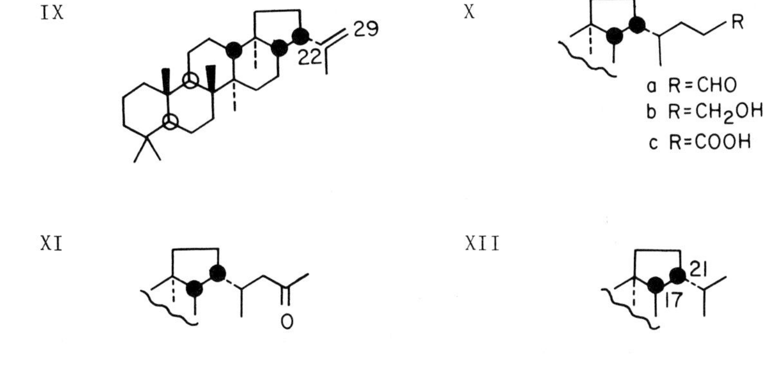

XIII XIV

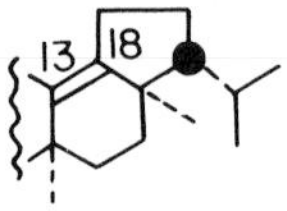

compounds have not been reported in organisms and are assumed to arise from a variety of transformations of functionalized precursors in microorganisms. Little is known about these transformations.

Partly as a result of the difficulties in synthesizing suitable radiolabelled hopanoids for incubation studies with surface sediments, no such experiments have yet been carried out. The only attempt at elucidating the early stage fate of any hopanoids (other than examining sedimentary distributions) has been a series of "decay" experiments. The first example is taken from a study of

XV XVI

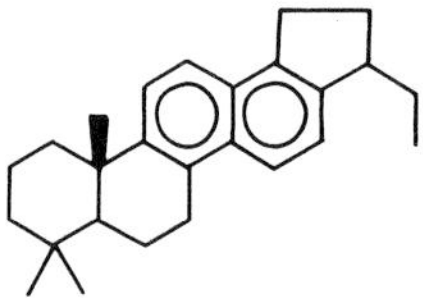

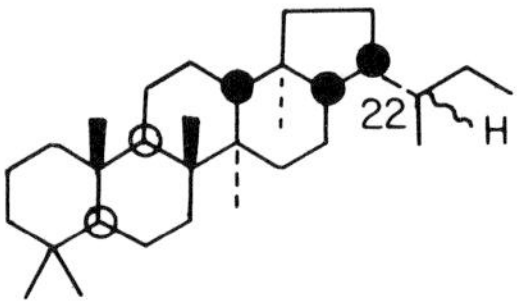

the hopanoid alkanes of the eutrophic lake, Rostherne Mere [see above; Quirk *et al.*, 1980*a*]. The alkanes from the following samples were examined: (i) a laboratory culture (non-axenic) of *M. aeruginosa*; (ii) a sample of predominantly *M. aeruginosa* collected from the lake; (iii) a similar sample allowed to "decay" on the lake surface and then on the lake bottom; (iv) the surface sediment (0 to 20 cm). Alkanes of the hopane type show an abundant ion at m/z 191 in their mass spectra and can be recognized readily in complex mixtures by mass fragmentography, using this ion. Figure 6 shows the m/z 191 fragmentograms from the gas chromatographic-mass spectrometric (GC-MS) analysis of the alkanes of the four samples. Compound H, 17β(H),21β(H)-hopane (XII) occurs widely in recent sediments [for example, Brooks *et al.*, 1977] and there is only one report of its occurrence in an organism [an acidophilic bacterium; De Rosa *et al.*, 1973]. Mass fragmentography revealed its virtual absence in the cultured sample (Fig. 6a). The two most abundant hopane hydrocarbons recognized were neohop-13(18)-ene (XIII; N in Fig. 6) in the alkane fraction (hindered double bond) and hop-22(29)-ene (IX) in the alkenes. The amount and concentration of 17β(H),21β(H)-hopane increased dramatically, however, in the two decayed samples. This is seen in Fig. 6, which contrasts the concentration of 17β(H), 21β(H)-hopane (XII) with that of neohop-13(18)-ene (XIII). In fact, the relative abundance of the two compounds in

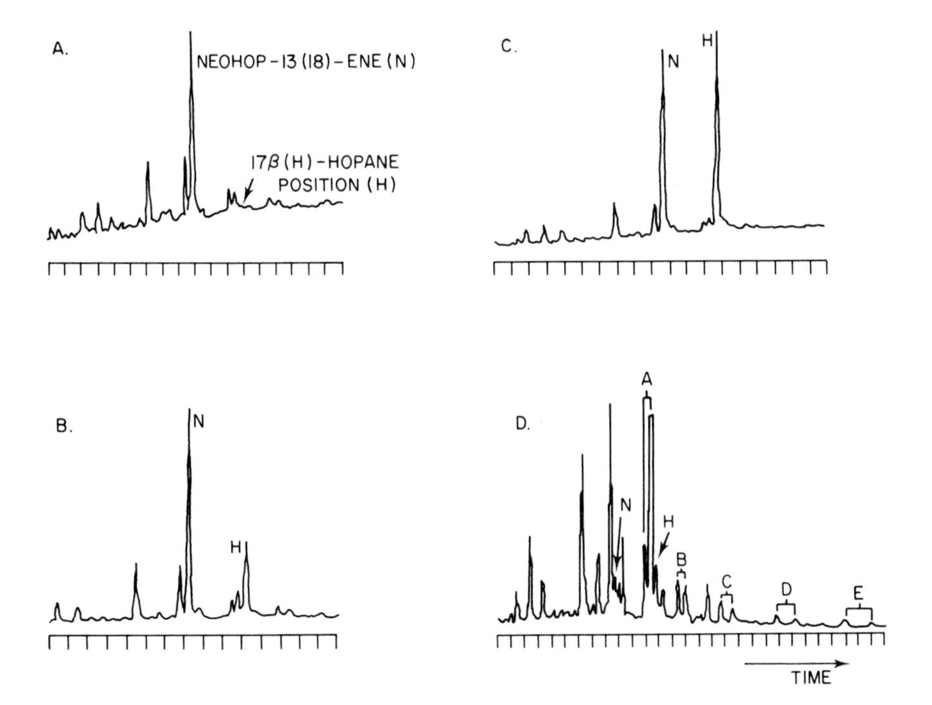

Fig. 6 Mass fragmentograms of m/z 191 from glass capillary GC-MS analyses of the branched and cyclic alkanes of four samples: **A.** *M. aeruginosa*, laboratory culture; **B.** *M. aeruginosa* collected from Rostherne Mere; **C.** *M. aeruginosa* decayed for 5 months on the lake surface (1 m depth) and 1 year on the lake bed; **D.** Rostherne Mere surface sediment (0 to 20 cm).

the sample decayed on the lake bed (Fig. 6c) is virtually the same as that in the sediment itself (Fig. 6d). This provides circumstantial evidence that this hopane originates from reduction of the abundant hop-22(29)-ene (IX) present. The culture sample, although not axenic, had no detectable *iso*- or *anteiso*-acids of bacterial origin, whereas the lake and lake bed samples had both present, suggesting high bacterial activity [Quirk, 1978]. It is possible, therefore, that the 17β(H),21β(H)-hopane arises from bacterial reduction of the alkene, hop-22(29)-ene. Components B to E (XIV, R = C_2H_5,n-C_3H_7,n-C_4H_9, and V respectively) in the sediment (Fig. 6d) have a distribution typically observed in petroleum and derive from a pollutant source [Dastillung and Albrecht, 1976]. This distribution arises from the effect of burial and the associated temperature rise, which isomerizes the biological configuration (17β(H),21β(H) and 22R; for example XVI) to the more thermodynamically stable configuration [17α(H),21β(H); Ensminger *et al.*, 1977] as a mixture (*ca.* 1:1) of the two

C-22 diastereoisomers for each carbon number (for example, XIV; see distribution in Fig. 6d). In the Rostherne sediment component A (XIV, R = CH_3) is an exception in that the latter eluting C-22 isomer is in much higher relative abundance. The major proportion by far of this isomer does not arise, therefore, from a pollutant origin and is an unusual example of a sedimentary hopane with the $17\alpha(H),21\beta(H)$ configuration, which does not arise from the effect of temperature. Although this isomer occurs widely in surface sediments [for example, Brooks *et al.*, 1977], it has not been isolated from any organism.

Further information about the appearance of this isomer was obtained by examination of the hopanoid alkanes of a predominantly sphagnum peat bog (Lyne of Skene; see above) [Quirk *et al.*, 1980*b*]. It is the major triterpenoid alkane in the peat, but was present in trace amounts in the living moss (component A in Fig. 7a) and even then in equal abundance to its C-22 isomer (showing both to be of pollutant origin). In the moss base (*ca.* 1 yr old), collected immediately below the living moss, a dramatic increase in its relative abundance was observed (Fig. 7b). This was also the case for a sample of moss decayed

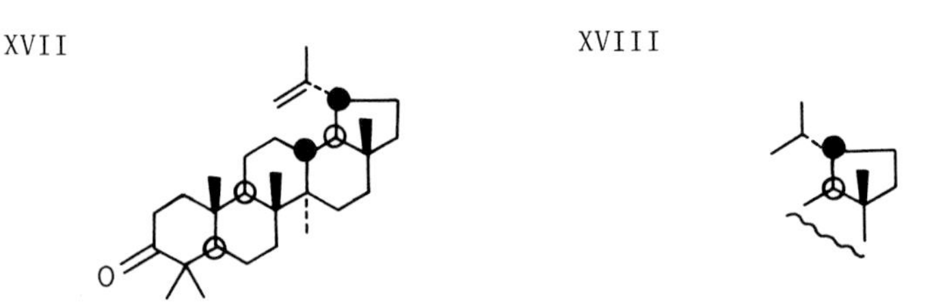

aerobically in the laboratory in the dark (Fig. 7c). The origin of this isomer in surface sediments appears, therefore, to be associated with bacterial activity. Whether it arises from transformation of a functionalized precursor in the moss, as yet unidentified, or is biosynthesised directly by the bacteria (mainly *Bacillus* spp.) is not known. Distinction between the two possibilities requires isolation of cultures from moss "decay" experiments and radiolabelling studies, and a detailed search in the moss for a likely precursor. The above studies emphasize the lack of knowledge about the early sedimentary transformations of hopane derivatives and the need for a combined microbiological and chemical approach.

Many surface sediments contain derivatives of non-hopanoid structure (examples include XIX-XXII), which arise from transformations of triterpenoids (for example, XVII, XVIII) contributed directly from higher plants. Two mechanisms have been put forward [Corbet *et al.*, 1980], whereby the products are formed by photolysis of the leaf ketones (XVII, XVIII) prior to incorporation into the sediment, or that they are photo-mimetic products formed by microorganisms through biochemical reactions

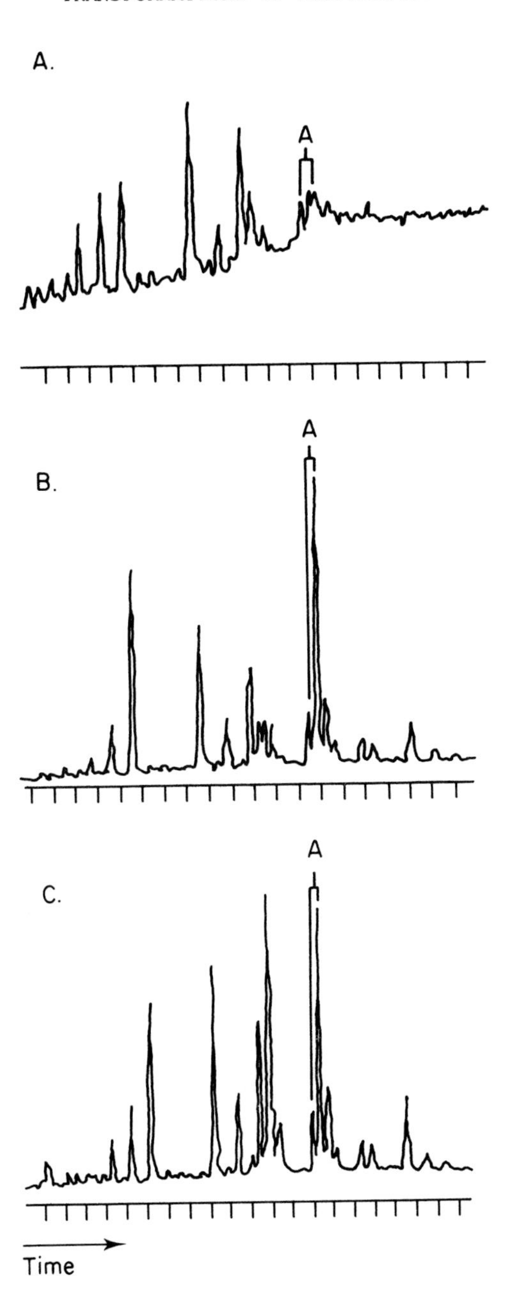

Fig. 7 Mass fragmentograms of m/z 191 from glass capillary GC-MS analyses of the branched and cyclic alkanes of three samples: **A.** *Sphagnum cuspidatum*; **B.** Moss base (*ca.* 1 year old); **C.** *Sphagnum cuspidatum* decayed in the laboratory (dark, aerobic, 30°C, 13 months).

XIX XX

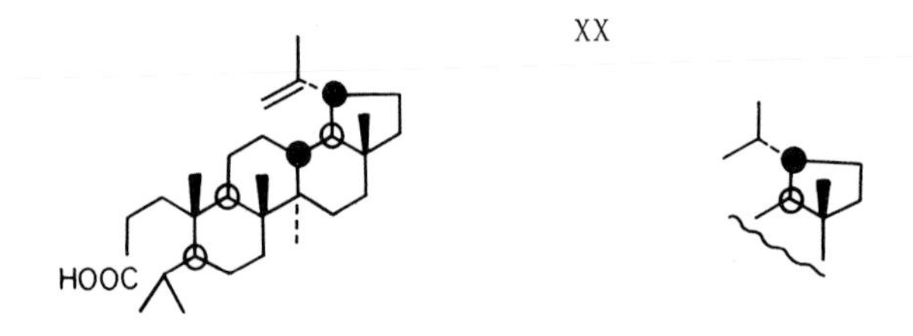

leading to the 3-ketones in their excited state prior to degradation.

Other triterpenoids derived from higher plant precursors and found in surface sediments are a series of aromatized compounds. These include components (XXIII-XXVI) arising from aromatization of β-amyrin (XXVII) or the corresponding 3-ketone, which are common higher plant triterpenoids. Those with four rings (XXIV, XXV) are probably formed by aromatization of hydrocarbons formed in the same way as components XXI and XXII. The finding

XXI XXII

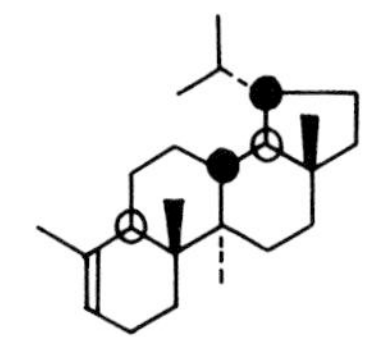

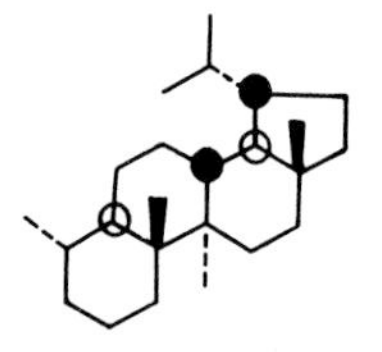

of these aromatic hydrocarbons in sediments only a few tens of years in age indicates that the aromatization is rapid and may be microbially mediated [Spykerelle *et al.*, 1977; Tissier and Spyckerelle, 1977; Tissier and Dastillung, 1978; Wakeham *et al.*, 1980], as it is unlikely that aromatization would occur by a solely chemical mechanism in these sediments. However, aromatization of a structure such as β-amyrin (XXVII) by microorganisms remains to be demonstrated.

XXIII XXIV

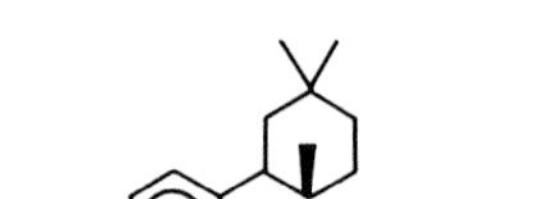

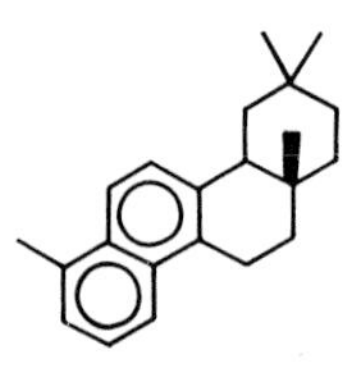

XXV XXVI

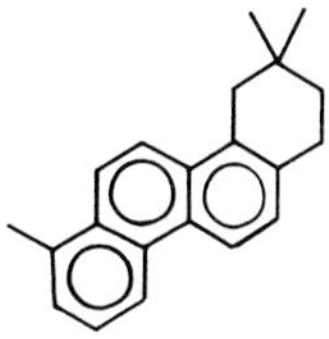

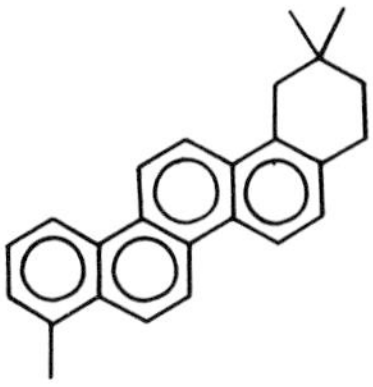

Acyclic Isoprenoids

Interest in the origin of such compounds arose from the fact that pristane (XXVIII) and phytane (XXIX) are two of the most abundant branched alkanes in sedimentary organic matter and their relative abundances to each other in sedimentary rocks and petroleum are used in exploration studies [for example, Clayton and Swetland, 1980]. It was

XXVII XXVIII

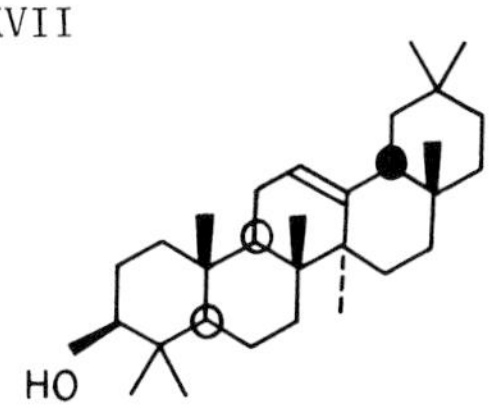

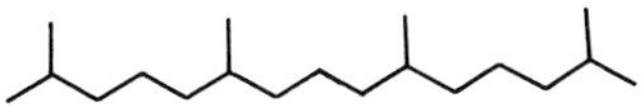

HO

originally suggested [Bendoraitis *et al.*, 1962] that they arise from the diterpenoid alcohol, phytol (XXXI, Fig. 8) which occurs widely in organisms, mainly esterified to the chlorophyll nucleus or to fatty acids. Correlation of the configurations of a suite of isoprenoid carboxylic acids ($\leq C_{20}$; for example VI) in an Eocene sedimentary rock with that of phytol provided further evidence that the alcohol is an important precursor of many sedimentary isoprenoids [Maxwell *et al.*, 1973]. Recent evidence indicates, however, that the lipids of archaebacteria may also be precursors of isoprenoid compounds thought previously to have a phytol origin [for example, Tornabene *et al.*, 1978].

This interest in the origins of sedimentary isoprenoid compounds led to studies of the early stage transformations of phytol in radiolabelling experiments. Incubations (*in situ* and in the laboratory) have been carried out in the surface sediment of Esthwaite Water, a eutrophic lake [Lake District, UK; Brooks and Maxwell, 1975; Brooks *et al.*, 1978; Quirk *et al.*, 1980*b*] and of Lonnekermeer, a shallow eutrophic pond [Eastern Netherlands; De Leeuw *et al.*, 1977]. The labelled products recognized are summarized in Fig. 8. The scheme is shown mainly to indicate the wide variety of transformation products which can be formed in sediments. It does not attempt to give any reaction sequences (for example, the C_{16} acid, XXXIV,

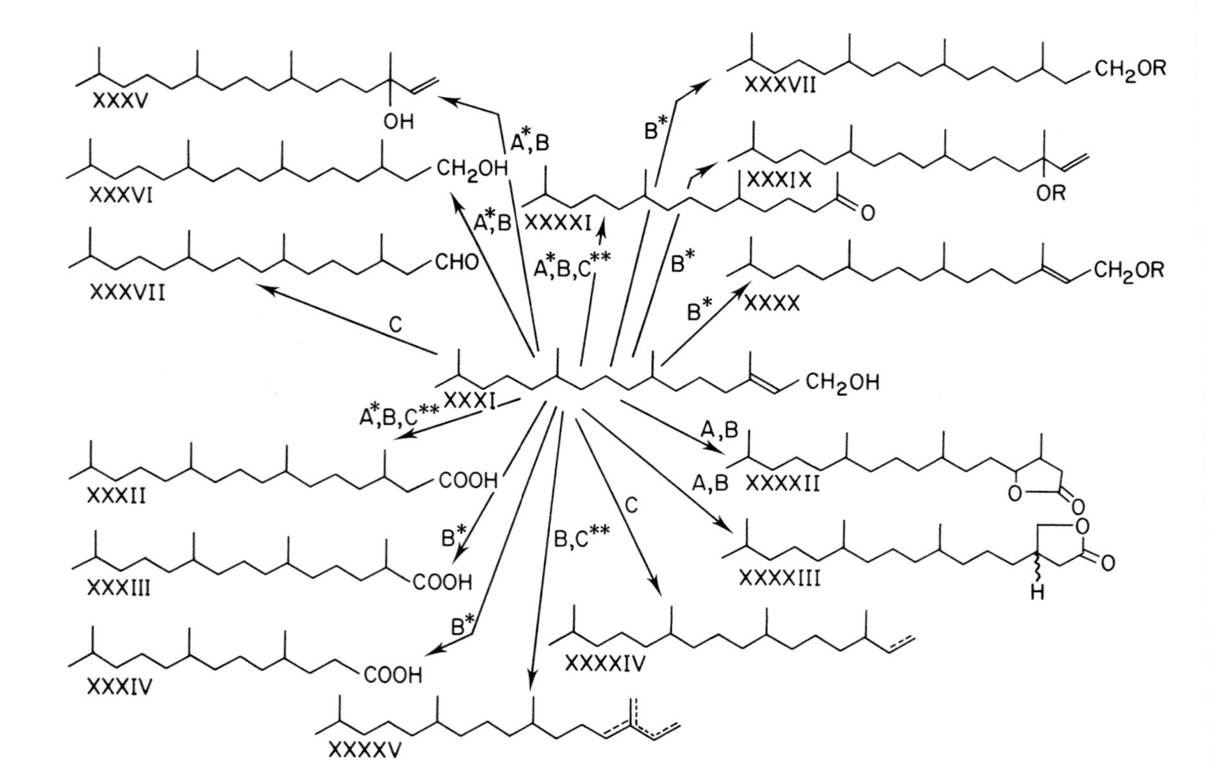

Fig. 8 Transformations of phytol in sediments inferred from incubation of U-^{14}C-phytol with: A) Esthwaite sediment *in situ*. B) Esthwaite sediment in the laboratory. C) Lonnekermeer sediment in the laboratory. * transformation product found in Esthwaite sediment. ** transformation product found in Lonnekermeer sediment.

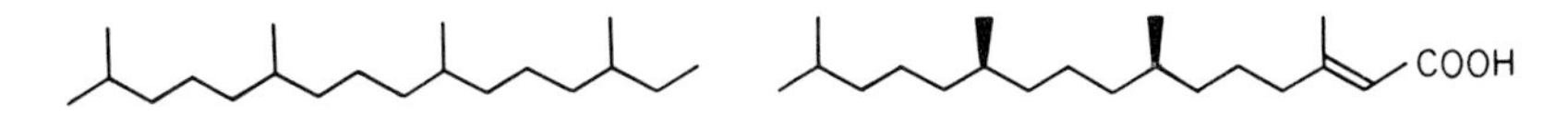

almost certainly arises from β-oxidation of the C_{19} acid, XXXIII). It also does not attempt to relate the differences in the products to changes in sedimentary environment, although these are likely to be important; De Leeuw *et al.* [1977] have suggested that the saturated C_{20} alcohol (XXXVI) is likely to be a product of anoxic conditions. The structures of most of the compounds show that they are biological products, although as before the extent of bacterial involvement is unknown. In a number of cases, analogous transformations have been observed in various biological systems.

Formation of the C_{20} acid (XXXII) from phytol (XXXI) has also been observed in the rumen of the cow [Patton and Benson, 1966], in bacteria cultured from Esthwaite Water sediment [phytol as sole carbon source; Brooks and Maxwell, 1975] in zooplankton intestines [Avigan and Blumer, 1968] and in animal tissue extracts [for example, Su and Schmid, 1975]. In the latter case the conversion has been shown to proceed *via* the unsaturated C_{20} acid (XXX), although the latter was not observed in the studies where bacteria are the presumed agents. Lonnekermeer sediment does contain this acid, but again it was not observed as a labelled incubation product [De Leeuw *et al.*, 1977]. The two lower molecular weight acids, XXXIII and XXXIV (Fig. 8), are known catabolic products (in animal tissue) of the C_{20} acid (XXXII), involving α- and β-oxidation steps [for example, Hutton and Steinberg, 1973]. Such transformations have not yet been demonstrated in a microbial system.

The saturated C_{20} alcohol (XXXVI) formed in the sediment incubations was also produced in the cow rumen study [Patton and Benson, 1966] and probably arises directly from microbial reduction of the phytol double bond. The rearrangement of phytol to XXXV occurs by an unknown mechanism. The role of the aldehyde (XXXVII) in the transformation scheme is also not known, although it may be a precursor of the saturated C_{20} acid. Elucidation of the various reaction sequences requires incubation with labelled intermediates such as this aldehyde, but no studies have been carried out.

Esterification of the alcohols (XXXI, XXXV, XXXVI) to form XXXVIII-XXXX is almost certainly brought about by enzymatic processes; the acid moieties of the esters consist primarily of n-C_{14} and n-C_{16} acids (corresponding to the R groups in Fig. 8) which are particularly abundant in bacteria [Kates, 1964]. Esterification of phytol has also been observed in thoracic duct lymph of rats [Baxter *et al.*, 1967].

Phytol can be oxidized to the C_{18} ketone (XXXXI) on standing in air, but such a mechanism is less likely in the incubation experiments under the reduced oxygen conditions in the sediments where the ketone was the major product. It was also a major product when a mixed culture of bacteria, cultured from Esthwaite surface sediment, were grown aerobically with phytol as sole carbon source [Brooks *et al.*, 1978] and may derive from enzymatic oxidative cleavage of the double bond *via* an epoxide [for example, Fonken and Johnson, 1972].

The lactones (XXXXII, XXXXIII) probably arise from hydroxylated acid intermediates, which cyclize readily to lactones. These products were not found in Esthwaite sediment itself, but were formed in the sediment bacterial culture experiments [Brooks *et al.*, 1978].

The origin of the dienes (XXXXV) is problematic; they were not detected in Esthwaite sediment itself and can arise easily from dehydration during the chromatographic procedures used for isolation. They were present, however, in trace quantities in the Lonnekermeer sediment [De Leeuw *et al.*, 1977] and have been found in the digestive tract of zooplankton [Blumer and Thomas, 1965]. The monoolefin (XXXXIV) is also of unknown origin, although such alkenes have been found in butterfat, where they are presumably derived from ingested phytol [Urbach and Stark, 1975].

Conclusions

Examination of the distributions of acyclic, tetracyclic and pentacyclic isoprenoids in surface sediments, limited radiolabelling incubation experiments, and comparison of the distributions with the natural products in organisms, indicate that a variety of transformations occurs in such sediments. In many cases, circumstantial evidence suggests that the reactions are brought about by bacteria, but the organisms responsible have never been isolated or identified. Collaboration between organic chemists and microbiologists is essential for understanding these early stage reactions on the degradative pathways leading to the derived compounds in sedimentary rocks and petroleums.

References

Albaiges, J. (1980). Fingerprinting petroleum pollutants in the Mediterranean Sea. In "Analytical Techniques in Environmental Chemistry" (Ed. J. Albaiges), pp.69-81. Pergamon Press, Oxford.

Avigan, J. and Blumer, M. (1968). On the origin of pristane in marine organisms. *Journal of Lipid Research* **9**, 350-352.

Baxter, J.H., Steinberg, D., Mize, C.E. and Avigan, J. (1967). Absorption and metabolism of uniformly ^{14}C-labelled phytol and phytanic acid by the intestine of the rat studied with thoracic duct cannulation. *Biochimica et Biophysica Acta* **137**, 277-290.

Bendoraitis, J.G., Brown, B.L. and Hepner, L.S. (1962). Isoprenoid hydrocarbons in petroleum. Isolation of 2,6,10,14-tetramethylpentadecane by high-temperature gas-liquid chromatography. *Analytical Chemistry* **34**, 49-53.

Björkhem, I. and Gustafsson, J.A. (1971). Mechanisms of microbial transformation of cholesterol into coprostanol. *European Journal of Biochemistry* **21**, 428-432.

Blumer, M. and Thomas, D.W. (1965). Phytadienes in zooplankton. *Science* **147**, 1148-1149.

Boon, J.J., De Leeuw, J.W. and Schenck, P.A. (1975). Organic geochemistry of Walvis Bay diatomaceous ooze - I. Occurrence and significance of the fatty acids. *Geochimica et Cosmochimica Acta* **39**, 1559-1565.

Boon, J.J., De Lange, F., Schuyl, P.J.W., De Leeuw, J.W. and Schenck, P.A. (1977). Organic geochemistry of Walvis Bay diatomaceous ooze - II. Occurrence and significance of the hydroxy fatty acids. In "Advances in Organic Geochemistry, 1975" (Eds. R. Campos and J. Goni), pp.255-272. ENADIMSA, Madrid.

Boon, J.J., De Leeuw, J.W. and Burlinghame, A.L. (1978). Organic geochemistry of Walvis Bay diatomaceous ooze - III. Structural analysis of the monoenoic and polycyclic fatty acids. *Geochimica et Cosmochimica Acta* **42**, 631-644.

Boon, J.J., Rijpstra, W.I.C., De Lange, F., De Leeuw, J.W., Yoshioka, M. and Schimizu, Y. (1979). The Black Sea sterol - a molecular fossil for dinoflagellate blooms. *Nature* **277**, 125-127.

Brooks, P.W. and Maxwell, J.R. (1975). Early stage fate of phytol in a recently deposited sediment. In "Advances in Organic Geochemistry 1973" (Eds. B. Tissot and F. Bienner), pp.977-991. Editions Technip, Paris.

Brooks, P.W., Eglinton, G., Gaskell, S.J., McHugh, D.J., Maxwell, J.R. and Philp, R.P. (1977). Lipids of recent sediments, Part II. Branched and cyclic alkanes and alkanoic acids of some temperate lacustrine and sub-tropical lagoonal/tidal-flat sediments. *Chemical Geology* **20**, 189-204.

Brooks, P.W., Maxwell, J.R. and Patience, R.L. (1978). Stereochemical relationships between phytol and phytanic acid, dihydrophytol and C_{18} ketone in recent sediments. *Geochimica et Cosmochimica Acta* **42**, 1175-1180.

Clayton, J.L. and Swetland, P.J. (1980). Petroleum generation and migration in Denver Basin. *Bulletin of the Association of American Petroleum Geologists* **64**, 1613-1633.

Corbet, B., Albrecht, P. and Ourisson, G. (1980). Photochemical or photomimetic fossil triterpenoids in sediments and petroleum. *Journal of the American Chemical Society* **102**, 1171-1173.

Cranwell, P.A. (1976). Decomposition of aquatic biota and sediment formation: organic compounds in detritus resulting from microbial attack on the alga *Ceratium hirundella*. *Freshwater Biology* **6**, 41-48.

Cranwell, P.A. (1979). Decomposition of aquatic biota and sediment formation: bound lipids in algal detritus and lake sediments. *Freshwater Biology* **9**, 305-313.

Dastillung, M. and Albrecht, P. (1976). Molecular test for oil pollution in surface sediments. *Marine Pollution Bulletin* **7**, 13-15.

Dastillung, M. and Albrecht, P. (1977). Δ^2-sterenes as diagenetic intermediates in sediments. *Nature* **269**, 678-679.

Dastillung, M., Albrecht, P. and Ourisson, G. (1980*a*). Aliphatic and polycyclic ketones in sediments. C_{27}-C_{35} ketones and aldehydes of the hopane series. *Journal of Chemical Research* **(M)**, 2325-2352; **(S)**, 166-167.

Dastillung, M., Albrecht, P. and Ourisson, G. (1980*b*). Aliphatic and polycyclic alcohols in sediments: hydroxylated derivatives of hopane and of 3-methylhopane. *Journal of Chemical Research* **(M)**, 2353-2374; **(S)**, 168-169.

De Rosa, M., Gambacorta, A., Minale, L. and Bu'Lock, J. (1973). Isoprenoids of *Bacillus acidocaldarius*. Phytochemistry **12**, 1117-1123.

De Leeuw, J.W., Simoneit, B.R., Boon, J.J., Irene, W., Rijpstra, C., De Lange, F., v.d. Leeden, J.C.W., Correia, V.A., Burlinghame, A.L. and Schenck, P.A. (1977). Phytol derived compounds in the geosphere. In "Advances in Organic Geochemistry 1975" (Eds. R. Campos and J. Goni), pp.61-79. ENADIMSA, Madrid.

Edmunds, K.L.H., Brassell, S.C. and Eglinton, G. (1980). The short-term diagenetic fate of 5α-cholestan-3β-ol: *in situ* radiolabelled incubation in algal mats. In "Advances in Organic Geochemistry 1979" (Eds. A.G. Douglas and J.R. Maxwell), pp.427-434. Pergamon Press, Oxford.

Ensminger, A., Albrecht, P., Ourisson, G. and Tissot, B. (1977). Evolution of polycyclic alkanes under the effect of burial (Early Toarcian shales, Paris Basin). In "Advances in Organic Geochemistry 1975" (Eds. R. Campos and J. Goni), pp.45-52. ENADIMSA, Madrid.

Ensminger, A., Joly, G. and Albrecht, P. (1978). Rearranged steranes in sediments and crude oils. *Tetrahedron Letters* **18**, 1575-1578.

Eyssen, H.J. and Parmentier, G.C. (1974). Biohydrogenation of sterols and fatty acids by intestinal microflora. *American Journal of Clinical Nutrition* **27**, 1329-1340.

Eyssen, H.J., Parmentier, G.C., Compernoll, F.C., De Pauw, G. and Piessenes-Denet, N. (1973). Biohydrogenation of sterols by *Eubacterium ATCC21,*408 Nova species. *European Journal of Biochemistry* **36**, 411-421.

Farrington, J.W. and Quinn, J.G. (1973). Biogeochemistry of fatty acids in recent sediments from Narragansett Bay, Rhode Island. *Geochimica et Cosmochimica Acta* **37**, 259-268.

Fonken, G.S. and Johnson, R.A. (1972). In "Chemical Oxidations with Microorganisms" Dekker, New York.

Gagosian, R.B. and Farrington, J.W. (1978). Sterenes in surface sediments from the southwest African shelf and slope. *Geochimica et Cosmochimica Acta* **42**, 1091-1101.

Gagosian, R.B., Smith, S.O., Lee, C., Farrington, J.W. and Frew, N.M. (1980). Steroid transformations in recent marine sediments. In "Advances in Organic Geochemistry 1979" (Eds. A.G. Douglas and J.R. Maxwell), pp.407-419. Pergamon Press, Oxford.

Gaskell, S.J. and Eglinton, G. (1975). Rapid hydrogenation of sterols in a contemporary lacustrine sediment. *Nature* **254**, 209-211.

Gaskell, S.J. and Eglinton, G. (1976). Sterols of a contemporary lacustrine sediment. *Geochimica et Cosmochimica Acta* **40**, 1221-1228.

Gaskell, S.J., Rhead, M.M., Brooks, P.W. and Eglinton, G. (1976). Diagenesis of oleic acid in an estuarine sediment. *Chemical Geology* **17**, 319-324.

Holland, H.L. and Diakow, P.R.P. (1979). Microbial hydroxylation of steroids. 5. Metabolism of androst-5-en-3,17-dione and related compounds by *Rhizopus arrhizus* ATCC 11145. *Canadian Journal of Chemistry* **57**, 436-440.

Hutton, D. and Steinberg, D. (1973). Identification of propionate as a degradation product of phytanic acid oxidation in rat and human tissues. *Journal of Biological Chemistry* **248**, 6871-6875.

Iizuka, H. and Naito, H. (1967). In "Microbial Transformations of Steroids and Alkaloids". University Park Press, State College, Pennsylvania.

Johns, R.B., Volkman, J.K. and Gillan, F.T. (1978). Kerogen precursors: chemical and biological alteration of lipids in the sedimentary surface layer. *The Australian Petroleum Exploration Association Journal* **18**, 157-160.

Johnson, R.W. and Calder, J.A. (1973). Early diagenesis of fatty acids and hydrocarbons in a salt marsh environment. *Geochimica et Cosmochimica Acta* **37**, 1943-1955.

Kates, M. (1964). Bacterial lipids. *Advances in Lipid Research* **2**, 17-90.

Kramli, A. and Horvath, J. (1948). Microbiological oxidation of sterols. *Nature* **162**, 619.

Lee, C., Farrington, J.W. and Gagosian, R.B. (1979). Sterol geochemistry of sediments from the western North Atlantic Ocean and adjacent coastal areas. *Geochimica et Cosmochimica Acta* **43**, 35-46.

Mackenzie, A.S., Patience, R.L., Maxwell, J.R., Vandenbroucke, M. and Durand, B. (1980). Molecular parameters of maturation in the Toarcian shales, Paris Basin, France - I. Changes in the configurations of acyclic isoprenoid alkanes, steranes and triterpanes. *Geochimica et Cosmochimica Acta* **44**, 1709-1721.

Maxwell, J.R., Cox, R.E., Eglinton, G., Pillinger, C.T., Ackman, R.G. and Hooper, S.N. (1973). Stereochemical studies of acyclic isoprenoid compounds. II. Role of chlorophyll in the derivation of isoprenoid-type acids in a lacustrine sediment. *Geochimica et Cosmochimica Acta* **37**, 297-313.

McEvoy, J., Eglinton, G. and Maxwell, J.R. (1981). Preliminary lipid analyses of sediments from Sections 467-3-3 and 467-97-2. In "Initial Reports of the Deep Sea Drilling Project" (Eds R.S. Yeats, B.U. Haq *et al.*), pp.763-764. U.S. Government Printing Office, Washington.

Mulheirn, L.J. and Ryback, G. (1975). Stereochemistry of some steranes from geological sources. *Nature* **256**, 301-302.

Nishimura, M. and Koyama, T. (1976). Stenols and stanols in lake sediments and diatoms. *Chemical Geology* **17**, 229-239.

Ogura, K. and Hanya, T. (1973). Cholestanol-cholesterol ratio in a 200-metre core sample of Lake Biwa. *Proceedings of the Japanese Academy* **49**, 201-204.

Ourisson, G., Albrecht, P. and Rohmer, M. (1979). The hopanoids. The palaeochemistry and biochemistry of a group of natural products. *Pure and Applied Chemistry* **51**, 709-729.

Patton, S. and Benson, A.A. (1966). Phytol metabolism in the bovine. *Biochimica et Biophysica Acta* **125**, 22-32.

Perry, G.J., Volkman, J.K., Johns, R.B. and Bavor, H.J. (1979). Fatty acids of bacterial origin in contemporary marine sediments. *Geochimica et Cosmochimica Acta* **43**, 1715-1725.

Quirk, M.M. (1978). Lipids of peats and lake environments. Ph.D. Thesis, University of Bristol, UK.

Quirk, M.M., Patience, R.L., Maxwell, J.R. and Wheatley, R.E. (1980*a*).

Recognition of the sources of isoprenoid alkanes in recent environments. In "Analytical Techniques in Environmental Chemistry" (Ed. J. Albaiges), pp.23-31. Pergamon Press, Oxford.

Quirk, M.M., Wardroper, A.M.K., Brooks, P.W., Wheatley, R.E. and Maxwell, J.R. (1980*b*). Transformations of acyclic and cyclic isoprenoids in recent sedimentary environments. In "Actes du Colloques" No.293 (Ed. R. Daumas), pp.225-232. C.N.R.S., Paris.

Rhead, M.M., Eglinton, G., Draffan, G.H. and England, P.J. (1971). Conversion of oleic acid to saturated fatty acids in Severn Estuary sediment. *Nature* **232**, 327-330.

Rhead, M.M., Eglinton, G. and England, P.J. (1972). Products of the short-term diagenesis of oleic acid in an estuarine sediment. In "Advances in Organic Geochemistry, 1971" (Eds. H.R. von Gaertner and H. Wehner), pp.305-315. Pergamon Press, Oxford.

Rosenfeld, R.S. and Gallagher, T.F. (1964). Further studies of the biotransformation of cholesterol to coprostanol. *Steroids* **4**, 515-520.

Rosenfeld, R.S. and Hellman, L. (1971). Reduction and esterification of cholesterol and β-sitosterol by homogenates of faeces. *Journal of Lipid Results* **12**, 192-197.

Rubinstein, I., Sieskind, O. and Albrecht, P. (1975). Rearranged sterenes in a shale: Occurrence and simulated formation. *Journal of the Chemical Society Perkin* **I**, 1833-1836.

Schubert, K. and Kaufmann, G. (1965). Bildung von Sterinestern in der Bakterienzelle. *Biochimica et Biophysica Acta* **106**, 592-597.

Seifert, W.K. and Moldowan, J.M. (1978). Applications of steranes, terpanes and monoaromatics to the maturation, migration and source of crude oils. *Geochimica et Cosmochimica Acta* **42**, 77-95.

Smith, A.G., Goodfellow, R. and Goad, L.J. (1972). The intermediacy of 3-oxo steroids in the conversion of cholest-5-en-3β-ol into 5α-cholestan-3β-ol by the starfish *Asteria rubens* and *Porania pulvillus*. *Biochemical Journal* **128**, 1371-1372.

Spyckerelle, C., Greiner, A.C., Albrecht, P. and Ourisson, G. (1977). Aromatic hydrocarbons from geological sources. III. A tetrahydrochrysene derived from triterpenes in recent and old sediments: 3,3,7-trimethyl-1,2,3,4-tetrahydrochrysene. *Journal of Chemical Research* **(M)**, 3746-3777; **(S)**, 330-331.

Stohs, S.J. and Haggerty, J.A. (1973). Metabolism of 4-cholestan-3-one to 5α-cholestan-3-one by leaf homogenates. *Phytochemistry* **12**, 2869-2872.

Su, K.L. and Schmid, H.O. (1975). Metabolism of long-chain isoprenoid alcohols. Incorporation of phytol and dihydrophytol into the lipid of rat brain. *Biochimica et Biophysica Acta* **380**, 119-126.

Tissier, M.J. and Dastillung, M. (1978). Inventaire et dynamique des lipides a l'interface eau de mer-sédiment. V. Hydrocarbures polyaromatiques des sédiments, de l'eau de mer et de l'eau interstitielle. In "Geochimie Organique des Sédiments Marins Profonds, Orgon II, Atlantique-N.E. Brésil", pp.275-283. C.N.R.S., Paris.

Tissier, M.J. and Spyckerelle, C. (1977). Hydrocarbures polyaromatiques des sédiments. In "Geochimie Organique des Sédiments Marins Profonds, Orgon I, Mer de Norvège", 229-236. C.N.R.S., Paris.

Tornabene, T.G., Wolfe, R.S., Balch, W.E., Holzer, G., Fox, R.E. and Oro, J. (1978). Phytanyl-glycerol ethers and squalenes in the archaebacterium *Methanobacterium thermoautotrophicum*. *Journal of Molecular Evolution* **11**, 259-266.

Urbach, G. and Stark, W. (1975). C-20 hydrocarbons of butter fat. *Journal of Agricultural Food Chemistry* **23**, 20-24.

Van Dorsselaer, A., Ensminger, A., Spyckerelle, C., Dastillung, M., Sieskind, O., Arpino, P., Albrecht, P., Ourisson, G., Brooks, P.W., Gaskell, S.J., Kimble, B.J., Philp, R.P., Maxwell, J.R. and Eglinton, G. (1974). Degraded and extended hopane derivatives (C_{27} to C_{35}) as ubiquitous geochemical markers. *Tetrahedron Letters* **14**, 1349-1352.

Wardroper, A.M.K., Maxwell, J.R. and Morris, R.J. (1978). Sterols of a diatomaceous ooze from Walvis Bay. *Steroids* **32**, 203-221.

Wakeham, S.G., Schaffner, C. and Giger, W. (1980). Polycyclic aromatic hydrocarbons in recent lake sediments - II. Compounds derived from biogenic precursors during early diagenesis. *Geochimica et Cosmochimica Acta* **44**, 415-429.

Withers, N.W., Tuttle, R.C., Holz, G.G., Beach, P.H., Goad, L.J. and Goodwin, T.W. (1978). Dehydrodinosterol, dinosterone and related sterols of a non-photosynthetic dinoflagellate. *Crypthecodinium cognii*. *Phytochemistry* **17**, 1987-1989.

SUBJECT INDEX